Multispecies Dialogues

AF842819

Multispecies Dialogues

Doing Philosophy with Animals, Children, the Sea and Others

Eva Meijer

Routledge
Taylor & Francis Group
LONDON AND NEW YORK

The publication of this book is made possible by a grant from the Nederlandse Organisatie voor Wetenschappelijk Onderzoek (NWO).

First published in 2025 by Amsterdam University Press Ltd.

Published 2025 by Routledge
4 Park Square, Milton Park, Abingdon, Oxon OX14 4RN
605 Third Avenue, New York, NY 10158

Routledge is an imprint of the Taylor & Francis Group, an informa business

© E. Meijer / Taylor & Francis Group 2025

The Open Access version of this book, available at www.taylorfrancis.com, has been made available under a Creative Commons Attribution-NonCommercial-NoDerivatives (CC-BY-NC-ND) 4.0 International license.

Trademark notice: Product or corporate names may be trademarks or registered trademarks, and are used only for identification and explanation without intent to infringe.

ISBN: 9789048564415 (hbk)
ISBN: 9781041183181 (pbk)
ISBN: 9781003700180 (ebk)
NUR 694

Cover illustration: Eva Meijer
Cover design: Coördesign, Leiden

DOI: 10.5117/9789048564415

Every effort has been made to obtain permission to use all copyrighted illustrations reproduced in this book. nonetheless, whosoever believes to have rights to this material is advised to contact the publisher.

For Product Safety Concerns and Information please contact our EU representative:
GPSR@taylorandfrancis.com
Taylor & Francis Verlag GmbH, Kaufingerstraße 24, 80331 München, Germany

For Olli and Doris

Table of Contents

1. All Thinking is Thinking with Others

Abstract: In the introduction I greet the reader of this book, and explain why dialogues are a good form for thinking with more-than-humans. I argue that we need to take language and dialogues more seriously in philosophy, and discuss how different genres of writing influence what can be discussed. I also discuss the concepts of 'world' and 'worldliness' and the importance of living differently for being able to think differently.

Keywords: multispecies dialogues, dialogical philosophy, worldliness, language, more-than-human philosophy

The first conversation is the one I have with you, reader. I write this text, you read it, and both writing and reading are intimate acts. Words can touch us in spoken and written form, and they can change us too. Thoughts and ideas are formed by the words we use, and how they move us is formed by our experience – our history, knowledge, way of being in the world. This is a book of philosophy, and philosophy is concerned with abstraction: in how it wants to clarify concepts and situations, or aims to shed light on reality by laying bare some of its structures. But it is also a book of dialogues, in which both I and others take part, a book that explores what language is through exploring what it does. In this text, my position as a writer and thinker is not neutral or distant: I am formed by the dialogues that I describe. Every 'I' comes into being through conversations with others – with family members, friends and companions of different species, as well as through conversations with larger society, such as those that take place in the public debate or through social media. This book aims to shed light on how dialogues shape us.

This is a very bad conversation, you might think now. I have not said anything, and do not even get the opportunity to introduce myself. You are right – this is a conversation with a very long delay between me writing the text and you reading it, and perhaps you will never speak back to me. But still, I cannot hold it on my own. I need you, the reader, to create the

Meijer, Eva. *Multispecies Dialogues: Doing Philosophy with Animals, Children, the Sea and Others.* Taylor & Francis Group, 2025
DOI: 10.5117/9789048564415_CH01

full meaning of this text. There is not one final meaning of this text that is mine to determine. In fact, this conversation we are having now is not one conversation, but a collection of conversations, with different readers who will all answer to these words in their own ways.

So, I invite you to follow me, reader, in thinking about what it means to speak with others, listen to them, and learn to recognize different types of conversations. In the chapters that follow I describe and analyze dialogues with nonhumans that are usually seen as incapable of conversations, such as mice, toads, plants, the sea, and art. And with fellow beings that should have more voice in the public and political debate, like nonhuman animals and human children. I do so in order to investigate what the concepts 'dialogue' and 'conversation' (I use these more or less interchangeably) mean and can mean, as well as investigate how we can become more worldly as humans. I write this in a time which is characterized by human domination, and violence towards nonhumans. This violence is very often interconnected with the refusal to recognize more-than-human others as thinking, feeling, and speaking beings. The conversations that follow offer the beginning of an alternative, which is imperfect, and far from finished, but still carries a promise – we can act differently.

Dialogues as a method for thinking with others

Dialogues have always played an important role in philosophy. In written form they were used by Plato when the language-game we call 'western philosophy' began. But conversations also play an important role in teaching, seminars, oral philosophical traditions, and many other practices. While conversations still make up an important part of philosophy, in teaching practices at universities, conferences or public talks, in written form dialogues have been replaced with other genres. In academic writing, 'text' now usually refers to rather specialized and abstract texts, written in generic English. In public philosophy, philosophical texts often come in the form of non-fiction books about how to live. This loss of dialogue has led to a loss of openness to the world in philosophy – it determines who takes part in philosophical conversations, how knowledge is created and for whom. How we write determines what we can say, and who can take part in the language-game in question. While dialogues do not automatically always do justice to all the voices involved, they do imply a focus on the other, or what is other in ourselves: a conversation needs different voices, questions and answers, a back and forth, an exchange, and it needs a location, because conversations take place in an actual space in an actual time.

My dialogues differ from Socrates' dialogues in significant ways. Because other-than-human beings take part in them, because their shape is different – they might include interventions in the landscape, or establishing daily rituals – and because their aim is different. I do not simply want to describe the dialogues and analyze them and, like Socrates, I am interested in a form of knowledge that lies behind what is immediately visible. But these dialogues are also an invitation to see other beings and language differently, and this book aims to offer the outlines of a framework for thinking and speaking better with others in a multispecies world, and not to find out a truth that is universal or lies outside of the conversation. In part, this is because the project of thinking with other animals has just begun in western philosophy and we do not know enough to make general claims about conversations with animals (Blattner et al. 2020). But in part this is also because it is not up to me, or any human, to define 'multispecies dialogues' at this point in time or draw general conclusions about other animals (Meijer 2019).

In my use of the concepts 'dialogue' and 'conversation' I implicitly (and sometimes explicitly, in Chapter 5) draw on Wittgenstein's ideas about language, specifically his ideas about language-games, family resemblance, and grammar (1958; Meijer 2019).[1] 'Dialogue' does not have one universal meaning that is the same in all cultures and ages; rather, we use the term to designate many practices that roughly share the same structure. Dialogues usually include questions and answers in some form, listening and speaking, multiple beings (even though we can also speak with ourselves, as I will discuss in Chapter 7). But their shape and function can be very different. A

1 Understanding that human language consists of many different language-games that create and express different forms of meaning and knowledge is important with regard to understanding how human language works too. A few years ago, I met a fellow philosopher at a conference, and we spoke about literature. He immediately began to speak about Elizabeth Costello, because this is what animal philosophers think about when they think about literature. His understanding of literature was different from mine (focused much more on 'story' and 'argument' than on language and ambiguity), and I felt a great distance from him. He then told me that there was probably some kind of communication possible with dogs, but that we could never understand dogs in the way we humans understood each other. But we do not understand each other at all, I thought, my dogs understand me much better than you do. The philosopher and I shared one specific language-game in detail, that of academic philosophy, while the dogs and I share many. Attentiveness to the many ways in which human language creates meaning can help us see that nonhumans also create meaning in many ways, and with learning to understand what they say.

good conversation with a friend or colleague can be thorough, and leave a strong mark on my thinking or feeling, while a conversation with a neighbor can be short and repetitive, and function simply to express friendliness. The long ongoing conversation I had with my dog companion Olli includes touch, while the conversations I have with human friends generally do not. Conversations with art and the sea do not involve human words at all, and challenge the limits of the concept of 'dialogue'. However, they might still share a family resemblance with what we currently see as conversation, and viewing interactions with nonhuman entities as dialogues can be of normative importance because it can shed light on how to interpret and do justice to nonhuman agency, as well as on how to work towards interspecies change.

In what follows I am not just interested in describing the dialogues that take place, but also in finding out what kind of dialogues humans could have with other beings. As Wittgenstein writes, there is always the possibility of new language-games (1958, I, 23). Humans in many parts of the world need new multispecies language-games – stories, political narratives, and conversations – to counter the violence I mentioned above and to develop more sustainable and caring ways of living. This also matters for philosophical reasons. We live in a time in which the 'grammar' about human and nonhuman beings is changing, in academia as well as the public debate. For example, in biology and ethology, animals (e.g. De Waal 2016), plants (e.g. Kimmerer 2013; Lawrence 2022; Marder 2013) and fungi (e.g. Adamatzky 2022) are discussed in very different terms than before: as active agents who affect humans in dynamic relations. In the humanities, scholars in the field of feminist political philosophy (e.g. Butler 2002; Young 1990), decolonial theory (e.g. Spivak 1988), and poststructuralism (e.g. Derrida 2008) have changed how we view 'the human', and how we understand the influence of power relations on concepts such as human and animal (Adams and Donovan 1995; Gaard 2016). In the political and public debate, discussions about the rights of nature have gained prominence in the past years (Tanasescu 2022), and animal parties are on the rise worldwide (Morini 2018). Climate change experiences affect the discourse about the relation between humans and nature, and voting preferences too (Hofmann et al. 2022).

These developments all challenge human exceptionalism, and ask us to view humans as part of multispecies worlds in which different beings exercise agency. Traditionally, philosophers have reserved concepts such as language, rationality, and agency solely for humans (Derrida 2008). Human exceptionalism in western philosophy is viscous. For example, even now, when the empirical evidence that nonhuman animals have complex forms of communication is overwhelming, there is very little attention for their voices and perspectives in academic thinking about moral and political questions (Meijer 2019). Prioritizing the perspectives of other beings is part of deconstructing anthropocentrism in theory and practice (Blattner et al. 2020; jones 2023). Rethinking human agency is too (Bennett 2020; Krause 2019). Dialogues offer a way to do both, and to rethink concepts like 'language' and 'reason' in the process.

Rethinking conversations from the ground up

Dialogues do justice to the agencies of different beings and offer us a chance to prioritize the perspectives of others, as well as to find new forms of interaction. Another reason to focus on dialogues as multispecies practice is that they already exist (Meijer 2019). This enables us to avoid idealistic forms of theorizing about others (Calarco 2015) and to connect theorizing to the existing realities of different beings (Haraway 2008).

Existing conversations between humans and nonhumans are currently often not recognized as conversations by humans, and they may include material and symbolic forms of silencing (Meijer 2022d) or other forms of violence (Wadiwel 2015, 2023). Analyzing these conversations and attending to nonhuman agency can show what is at stake for different interlocutors and offer an alternative. Conversations are connected to transformation too. If we seriously engage in conversation, within oneself, in our private lives, or in relation to political others or between communities, we run the risk of being changed (Habermas 1994, Young 2002). However, in order for actual change to take place, the different beings who speak in the dialogue need to be open to others. Without interest in the other, turn-taking, questions, and listening (Bickford 1996) dialogues turn into monologues (Meijer 2022c).

The fact that we are constantly in conversation with others should also be made more explicit in philosophy and other forms of knowledge creation more generally, and philosophers should be more aware of their environment, including their interlocutors, in analyzing the world. In academic philosophy, and perhaps written philosophy in general, it is now the case that a specific type of human aims to give a truthful picture of a certain phenomenon, which structures the larger conversation. The language-game of philosophy carries a strong stamp of a certain kind of human (one that likes thinking and arguments, a certain form of distance and abstraction, and so on), because how we are situated determines how and what we think. This affects the theories that are being developed, and how they are discussed. Those working at universities also often belong to a certain class and opportunities for a career in academia are connected to factors like gender (Asenbaum 2022).

Focusing on democratic theory, Hans Asenbaum (2022) points out that this way of working leads to a tension. Democratic theory, in the past decades, has been concerned with theorizing inclusion, agency, and transparency, while it lacked these qualities in the process of theorizing. Asenbaum writes that we need to take thinking with others seriously for democratic reasons, and that theory can function as a democratic innovation. In order for theory to be more democratic, the humans (or nonhumans) that are the object of theorizing should, for example, have a voice in the process of theorizing, and the author of a text should understand herself as part of the assemblage that generates knowledge.

Writing from the first-person perspective

In the dialogues that follow I often take part. The 'I' in this text is a lyrical I, and this book is not a memoir, nor is it autobiographical – the focus is not on giving a truthful picture of my life but on the conversations. Still, my experience is part of the investigations. This is necessary to do justice to my interlocutors, to the sort of dialogues I am describing, and to acknowledge that this text and I are shaped by more agencies than just mine. The fact that there is an I in the story does not make it less critical than disembodied forms of writing philosophy; including one's own position when discussing interaction with others can clarify power structures. Another benefit of this approach for multispecies thinking is that it situates the text in the actual world, which matters for doing justice to the perspectives of beings who do not write or read.

In the history of philosophy, philosophers used different genres and ways of writing to better understand life and the world, including dialogues, confessions, manifestos, remarks, autobiographies, and even novels. The genre that you choose matters not only for the meaning but also for the worldliness of a text. Think, for example, of the difference between Descartes (2013) and Montaigne (2019). Descartes and Montaigne had different views about what method to use for thinking, and used language differently to express their views, which shaped how they portrayed reality (Adorno 1984) and their relation to other beings, specifically more-than-human animals (Melehy 2006). In his *Meditations* (2013), Descartes begins to speak from his own perspective as a philosopher on a chair, but soon turns inward and focuses solely on thinking, while he strips away all contingencies in search of a universal truth. He emphasized that anyone can follow him, if they take the same steps in their minds. In contrast, Montaigne (2019) takes a phenomenological approach, and he himself features in his essays as a living being, who withdraws from the world into the tower of his castle in order to write (we probably cannot escape that one step back when writing: my writing begins in silence too), but never leaves that world. He takes his own experience – his life, his body, his relations – as a starting point for thinking and situates his ideas in a world with real wars, real humans and real cats. His essays also take seriously the lives and lived experiences of others, and combine different genres: he quotes freely from what he reads, often referring to the ancient Greeks and Romans, but also from his contemporaries. Poetry has as much right to speak as philosophy.

Montaigne and Descartes are on opposite sides of the spectrum when it comes to taking seriously nonhuman animals, and this divide is connected to their attitude towards text, knowledge, and philosophy. 'Man is the weakest and most vulnerable of all creatures, and at the same time the most arrogant,' writes Montaigne in his *Apology for Raymond Sebond* (2004, 530), an essay about nonhuman animals. This sentence is followed by a series of stories about the lives of the other animals with whom humans share the world, and who according to Montaigne have their own forms of wisdom, culture, and language. There is much that they can do better than humans, Montaigne emphasizes – for example building nests, being loyal, or understanding each other – and there is much in which we are not so different. There are also many ways of understanding one another. If humans do not understand certain animals, it is not because of a lack on their side, but on ours. Moreover, there are similar problems with understanding unfamiliar groups of humans. In this essay he gives many examples of understandings and misunderstanding between animals including humans, also from his own life. The most well-known example concerns his cat companion. When he is playing with this cat, he writes, it is unclear who is playing with whom. This commonsense attribution of agency to another animal is not radical in daily discourse, but it has become so in philosophy.

Acknowledging the inner lives and agency of nonhuman animals became radical in part because of the legacy of Descartes, who draws different lines than Montaigne. Reason is a capacity of humans, he writes, which is connected to their immortal soul (Descartes, 1637, 42; Derrida 2008). Nonhuman animals, which he calls bêtes-machines, function like clocks, mechanical beings that react but do not feel because they do not think (Thomas 2020), and we know that because they do not speak (Descartes

1637, 42). There are animals, like magpies, who may learn to speak in human language, but that is only imitation of human sound: they cannot take part in conversations (Descartes 1637, 42). Derrida writes that Descartes does not really address the question of whether animals can think; rather, he reformulates the question in such a way that they can only react and not respond (2008, 83). In the work of Montaigne, animals do answer, and even ask questions (1958, 334).

Of course, Descartes is also an 'I' in a text, who lived in a certain time, sat by the fire, and described his own sensations. But in his thinking, he tries to move beyond this contingent I, in order to be able to capture the truth. In an essay about the essay, Adorno problematizes this attitude towards the truth and connects it to Descartes' style (1984). Adorno writes that there is a tendency in philosophy to want to be complete, to write a text that is always and universally true. But by doing so, you do injustice to reality. Reality cannot be caught, writes Adorno; that would be an act of violence. He proposes the essay as a more suitable form for approaching the world, because the essay shows phenomena instead of proving them: it illuminates them, does not try to catch them and pin them down. An essay is simultaneously more open and closed than a philosophical text, and in the essay there is no first principle and there are no ultimate principles.

Adorno's essay raises many questions, including questions about the agency of the writer and that of the work of art, a topic that I will turn to in Chapter 6. For now, I simply want to draw attention to the fact that the genre we choose determines how we approach the world, and that seemingly neutral forms of defining, which are popular in current academic philosophical practices, are not neutral and may even obscure how the author is situated, which affects the knowledge that is presented to the reader (Haraway 1991). In this context, Adorno warns us for 'instrumental reason', a form of human rationality that objectifies the world, including humans and human relations, and that reads reality in terms of efficiency and rationality. In the sort of reasoning that Descartes employs, Adorno argues, the tendency towards identity and unity, which is found in much human thought, suppresses difference and diversity (see also Derrida 2008 on the interconnections between how the concepts '*logos*' and 'animal' are defined in the philosophical tradition). Things that are not actually equal are made equal, and given an exchange value. Instead, Adorno writes, the essay does justice to the difference, what is other, the intangible, because it precisely walks alongside what is unique without capturing it. This walking-with, or aligning oneself with, shares similarities with what Jane Bennett (2020, xx) calls 'writing up', a form of using language that is more flexible than common academic discourse, and recognizes how our writing practices are influenced (disrupted, enabled, changed, moved) by outside human and nonhuman forces. Mapping more-than-human dialogues asks for attending to the way in which we use language, and doing justice to the agencies of others does too.

Here I want to note that while this book is a theoretical endeavor, the theory is interconnected with practice. With the exception of Chapter 5 and part of Chapter 6, the conversations that follow are actual conversations that I had or am having with actual others. Exploring the model of the dialogue for reformulating relations is not simply a philosophical interest for me, it is part of learning to live differently. Living differently is necessary for being able to think differently.

Speaking with others

The connection between learning to live differently with others and speaking differently with others is most clear in the first three chapters, in which I describe and analyze conversations with my dog companion Olli, a group of ex-laboratory mice I adopted, and migrating frogs, toads and salamanders.

The conversations I had with Olli and describe in Chapter 2 changed my life and the direction of my work about animal agency and the multispecies community. As a dog who grew up on the streets of Romania, Olli was his own person before he came to live with me, and he co-shaped my views about dogs and working towards more just multispecies societies with other animals. He also helped me think about the similarities and differences in human-human and dog-human language-games. The first part of the chapter is devoted to the role of touch in multispecies language-games, because this was one of the main ways in which Olli shaped our conversations. Zooming in on the role of touch in our conversations shows how new multispecies language-games can come into being, if we begin to pay attention to others. In the second part of the chapter I investigate how we can take dog agency seriously politically, through analyzing Olli's diplomatic skills, which are much better than mine. Olli not only helped me to understand the kind of wisdom that dogs have, but also showed me how this can function as the basis for new social and political multispecies action.

In Chapter 3 I discuss a rather different set of conversations which I had with a group of ten female laboratory mice who came to live with me in the summer of 2020. I had little experience with mice when these people came to live with me, and was immediately struck by their practices of care, their friendships and their individuality. By looking at them over a long period of time, I learned to see them and understand basic aspects of their language. In the conversations with Olli, touch was important. While touch is an important part of the communication between the mice, they did not want to be touched by me. In our conversations sound and music played an important role, as well as material interventions and the creation of new habits. In the chapter I not only describe how the conversations between the mice and me developed, but also discuss the politics of living with mice, and possibilities for deliberation.

The material and spatial dimensions of communication – where we speak with others and how – also play an important role in the conversations I discuss in Chapter 4, with amphibian and human neighbors in my town. In the same year that I adopted the mice, I moved out of the city to a watery town that is a home to many frogs, toads and salamanders. These animals hibernate in the gardens in winter and move to the ponds in early spring. Because they are still slow when they wake up, they need assistance to cross the streets; otherwise they get hit by cars or buses. In the first year, I witnessed many casualties and decided to found a working group to help them, *Paddenwerkgroep Landsmeer*. In the chapter I describe how my experience of the town, the weather and the seasons, and that of the other volunteers, changed through engaging with these amphibians. I also discuss what this can tell us about the embodied and spatial character of conversations more generally, and how conversations with amphibians relate to, and influence, dialogues between human neighbors, and the narrative of the town.

In Chapters 5 and 6 I further investigate speaking with nonhumans. I do so in order to learn to live better with them, but also to explore the limits of the concepts 'dialogue' and 'conversation'. In Chapter 5 I focus on conversations about and with the North Sea. I compare dialogues with (sea) animals to those humans have with plants, epistemologically and normatively, and argue that more-than-human beings should have a right to speak about our common world. I also explore how we can engage differently with the North Sea itself by looking at existing examples of humans who speak with the sea. Viewing interaction with the sea as conversations can be of normative importance and change the attitude of humans. It also directs our view as to how human language works and asks us to be attentive to the varied meanings that concepts can have in different situations.

Speaking with the sea asks for acknowledging that human agency is always part of a larger network of relations, in which human and nonhuman beings exercise different forms of agency. To further explore this, I turn to the conversation I have with my work in Chapter 6, using writing novels as an example. The ongoing dialogue I have with my work forms my days and character, but in a different way than the conversations I have with Olli or other animals. I argue that the porousness that is required to write shows us something about human agency more generally, and also matters democratically. Drawing on Bonnie Honig's ideas about public things, I investigate the role of art in social and political conversations between humans. I also discuss how art can create meaning in multispecies relations.

There is one dialogue that is woven through all of the others, and that is the dialogue that I have with myself. In Chapter 7 I discuss the dialogue I have had with myself about depression in the past thirty years. I contrast this dialogue with Hannah Arendt's view of thinking as a dialogue between 'I and I', in order to develop a more embodied and situated view of speaking with oneself. I also investigate how conversations with the self are related to those with others, in the context of depression. I analyze how (historical) power structures affect how we collectively think about depression, and how the (medical, political, existential) lens that we use to describe and treat depression, affects how we can speak with ourselves about depression. I then turn to the importance of living well with others for living well with oneself in a violent world. In the last part of the chapter I discuss how viewing thinking as a social and embodied practice can inform new forms of multispecies public philosophy.

In Chapter 8 I further discuss the question of learning to live well with others, which in our time asks for multispecies education. Developing a better understanding of multispecies education, and new practices, is not something that can be done by human adults from behind their desks. This asks for practice-based learning with nonhuman animals and plants, and for including the perspectives of human children. In the chapter I discuss questions about multispecies education and learning to live better with others more generally with two groups of children.

The dialogues in this book do not only include different forms of language, speaking and expression: they also include silence, and listening. As a conclusion, in Chapter 9 I discuss how we can learn to have better conversations with others through an exploration of the role of silence and listening in multispecies politics. While certain forms of – non-invasive – ethological and biological research, as well as artistic experiment, can play a role in improving interactions with nonhumans such as animals and plants, the most important task for humans in our age is to take a step back and listen. We should invite others to speak in existing debates and deliberation about our common life-worlds and habitats, but also let them determine if they want to speak with us at all, and when they do, allow them to (co-)create the conditions under which we speak. Learning to speak better with others – human or nonhuman – begins with listening. For individuals, but also for societies. Learning to listen is not just an ethical, but also a political task.

From world to worlds

In the beginning of this introduction, I wrote that one of the aims of this book is to investigate how we can become more worldly as humans. I already mentioned some possible ways in which we can go about this task – we can think with others and take their perspectives seriously, reconsider the language we use and how we speak, and create knowledge with others – and I will address the question of worldliness at several points in the next chapters. But before we get started, I need to elaborate briefly on how I use the concept world, because I take it to have different meanings that matter for what follows.

When I speak of 'becoming more worldly', I appeal to worldliness as a virtue, something that humans need to (re)learn in other to be able to interact in more just and attentive ways with members of their own and other species. Part of becoming more worldly is building new worlds, or strengthening existing worlds with others, by which I mean common life-worlds, social worlds that already are multispecies worlds (Donaldson and Kymlicka 2011; Emel and Nirmal 2021; Haraway 2008), in which humans are animals too (Srinivasan 2022). I also use 'world' to refer to the life-worlds of others, in the sense that anthropologist Arturo Escobar proposes when he argues for the pluriverse (2020). Escobar writes that different human communities not only have different styles of living but also different knowledge systems: worlds consist not just of practices, but have their own cosmological systems. Recognizing the plurality in these worlds is not an obstacle for living more justly with others, but an opportunity, Escobar writes (2020). Those in the west can learn from indigenous communities. To this I would add that these other cosmologies are not solely human: animal and plant communities have their own forms of wisdom.

Pluralizing the notion of world – recognizing the concept's different meanings and understanding that there are different life-worlds on planet Earth – matters for several reasons. Pragmatically, because in the climate crisis one of the prominent narratives is that we are losing the world, planet Earth. While this may or may not be true, speaking of 'losing the world' obscures what is at stake, i.e. the many different life-worlds that exist, of which only some matter to humans, and that have value for their own sake. This can paralyze humans who do want to do better because it makes it seem as if individual acts do not matter in the larger context (McKinnon 2014).

Moving from world to worlds also matters politically: different groups of human and nonhuman others may perceive the world differently and have different forms of agency due to power relations (Emel and Nirmal 2021). Differences generally do not necessarily obstruct understanding, as they give us something to speak about, but standpoints and power structures govern the shape of our agency and position in common life-worlds. There are also many worlds that are already lost, human and more-than-human, because of capitalist, colonial and anthropocentric exploitation, which should inform political considerations.

Finally, recognizing that there are different worlds matters philosophically. For epistemic reasons, because different beings have their own perspectives on life, and attending to these provides us with a better picture of reality (Blattner et al. 2020), and for moral reasons, as these other perspectives and voices count too. In thinking through questions about sharing space or solutions for ecological problems, the perspectives of more-than-human beings like nonhuman animals and plants should be foregrounded because they have been silenced for so long (Meijer 2019; Tschakert et al. 2021). Understanding that there are different worlds that matter and different opportunities for forming new life-worlds with others can help direct multispecies 'worlding' (Haraway 2016, 13). Multispecies worlding may entail activism or philosophical analysis (Celermajer et al. 2020; Emel and Nirmal 2021), but it can also involve other practices such as planting trees, taking care of green areas in cities, multispecies education (Editorial Team 2023), conducting public debates carefully, designing a politics that is focused on care and healing instead of conflict (Adams and Donovan 2007), or assisting other animals in rebuilding their lives and communities (jones 2019).

I know, reader, that at this point, the dialogue between us is beginning to sound like a monologue. But it is nearly time for the other voices. There is one more thing I want to say, and that is that dialogues hold a promise, and that we can learn to speak better with others. Entering into a serious dialogue with someone else is demanding, and it is risky, because there is always the possibility that it might change you. There never is a guarantee that a dialogue will lead to understanding or a better situation for the different beings involved, but something new can come out of them. And as humans, we can learn to speak and live better with others. The following dialogues are exercises in doing so.

CONVERSATIONS WITH DOGS

I

Hello, says Olli by touching my hand very softly with his nose.

'Hello friend,' I say with my voice. With my hand I stroke the side of his neck. Olli is old and his sight and hearing are not good. 'Are you hungry?'

I get up because this greeting means he is indeed hungry – he does not like to eat much in the morning anymore, but then gets hungry a couple of hours later. He follows me to the hallway, where I open the cupboard and get out some bread. I let him smell it. He says no by turning his head away. I take the next few steps into the kitchen and offer him a lick of peanut butter.

No.

I open the fridge, and offer him some leftover pasta. He takes a bite, so I fill his bowl.

Olli eats the pasta and then comes up to me again to thank me, like he has thanked me for every meal that I gave him in the past ten years.

II

Doris sits down on the carpet next to the table in our friend B.'s house. It took her some time to find a spot, because she is scared of the floor – she finds it too slippery.

'You can lie down if you want to,' I tell her.

She looks up. A large plant is towering over her head.

'Sorry,' says B., and she moves the plant to the side so that Doris has free space around her, and feels safe enough to lie down.

III

Doris is in Olli's bed near the window. Earlier, he was asleep on the couch. When he returns from the garden, he steps into his bed, nearly on top of her – he did not see her, because his sight is bad. She gives me a look of total surprise, and from that moment on understands that he is different from before.

IV

When Olli had just arrived from Romania, I made him wear a harness and collar. He soon found out that this not only restricted him, but gave him power over me too. In the walks that followed, he simply stopped if he did not want to go somewhere. Sometimes he lay down, sometimes he pulled in the other direction. Sometimes I made him walk (should I have? I always meant well), sometimes I followed him, sometimes we found another solution. If he walked off leash – in parks, and when we moved to a small town on most walks – we just followed one another.

2. Conversations with Olli

Abstract: In this chapter I describe the dialogues I had with the Romanian street dog, Olli. In the first half of the chapter, I focus specifically on the role of touch in dog-human language-games, drawing on insights about touch in phenomenology, critical animal studies, and the work of Irigaray. I then turn to animal politics, and investigate how we can take dog agency seriously politically, through an investigation of Olli's diplomatic skills. Olli not only helped me to understand the kind of wisdom and insight that dogs have with regard to the multispecies community, but also showed me how this understanding can function as the basis for new social and political multispecies action.

Keywords: animal politics, dog philosophy, auto-ethnography, touch, animal dialogues, multispecies community

At the train station in Pașcani, common house martins had built three nests in an unused window frame.[1] The parents flew in and out of the clay constructions with insects. I heard the young birds call for food when I had left the platform and stood there for a while, watching them. It felt like a good sign.

When the martins flew away to find more food, I followed the stream of humans who just got off the train into town. Some went left, others right – I chose to go right. I had a hunch that this was the right way. The road led up a hill, past a market – most stalls sold flowers and plants, tulips were popular – and a large church. The window sills in the church tower were occupied by pigeons, who watched what went on below.

When I was halfway up the hill, I took a turn to the left. The first dog I saw – blonde and sturdy – was locked in a garden. Maybe her job was to watch the house, but she only wanted contact. Two other dogs were small

1 The introduction of this chapter was published as 'A Stranger in your Town' in a modified form in *Leonora Magazine*, forthcoming.

Meijer, Eva. *Multispecies Dialogues: Doing Philosophy with Animals, Children, the Sea and Others.* Taylor & Francis Group, 2025
DOI: 10.5117/9789048564415_CH02

enough to enter and leave the gardens through holes in the fences. They seemed to be looking for something and had no interest in me. I reached the upper part of the city center and looked out over the town before I took the stairs down to the lower part of town. Dogs had made small paths through the high grass on both sides of the stairs. In the church near the square, men were singing hymns, the sound of their voices reached far into the streets. When I sat down on one of the benches in the main street, a very beautiful young dog walked up to the fountain and lay down in the shade. Nobody seemed to notice her.

My dog companion Olli was born in Pașcani. He lived on the streets of this town for several years. When he was three or four years old, regulations regarding street dogs changed, and in Pașcani, like many other Romanian towns, dogs were caught by dog catchers and taken to shelters. A local animal activist told me that Romanian humans generally do not adopt dogs from shelters, so the dogs there are sentenced to a life in captivity, separated from their companions and friends. Olli ended up in the municipal shelter, where the dogs did not get enough food nor medical care, and were treated violently. Many of them died. A small Dutch organization brought him to another shelter, and a few months later I found his profile on their website.

When Olli arrived in Amsterdam, everything was new: the scents, the city, living in a house, living in close proximity to a human and a cat (Meijer 2014). My dog companion Pika helped him to find his feet, and Olli was eager to learn. In the first months together he and I developed a common language, which included human words, dog and human touch, growls (often used by him as an invitation), and deliberation about the leash. We developed common habits. My days were shaped by him, and my outlook on life too. He was not a puppy who needed education nor an unruly dog who needed training, but an adult with his own perspective on life, and his own moral compass.

At the time I was writing my PhD thesis about political animal voices (Meijer 2019) and forming new multispecies communities, and Olli helped me to form my views. Street animals show us how new multispecies relations can look, relations in which humans do not dominate the other animals, but share spaces with them on the basis of freedom and trust. He also taught me about the importance of beginning again. He was so afraid when we met, and managed to turn this around, always treating everyone with the greatest gentleness and kindness.

Because I wrote about us getting to know each other, I felt I understood the transition that Olli had to make, from his first life to his second life. But walking in Pașcani, I understood something else about it, because now

I was the stranger in his town. The roads, the gardens full of flowers, the church, the shops, the humans – it all seemed so unfamiliar.

I looked at the map, and decided to walk to the other side of town, with the large park, and then take the road next to the railroads back into the city. My aim was to find out where and how the street dogs lived. In the days before I had been in Iași, a larger town nearby, to give talks at the university. As in Pașcani, in Iași there were no large groups of street dogs, I only saw individuals and couples. When the dogs were on the move, they moved fast – Olli did the same when he came to Amsterdam, and because this was his preferred way of moving, I took him on runs. The dogs also treated humans in a specific way. They ignored most of them, but some familiar humans they approached, wagging their tail enthusiastically, very clearly showing their politeness and friendliness. Olli did this too in the first months of our relationship, and when he stopped performing this dance, I felt he began to trust me. The dogs who rested in public areas remained on guard. When Olli arrived, it took him months to start sleeping deeply, and when cat Putih or I disturbed him accidentally when he was sleeping in the first year, he jumped up. He also had nightmares.

The sun was high in the sky when I resumed my walk. I followed two smaller dogs through the flowerbeds between the apartment buildings just off the main street. They went up to a house and began whining in front of a yellow door. It was probably dinner time. When the door remained closed, they lay down not far from the house and waited. A man on a bench was watching me watch the dogs. Just outside of town, near the large park, two larger brown dogs had a heated conversation – the female was barking and sniffing around, it looked like an intruder had come to visit. Here Olli could have lived. I could not see him living near the apartment buildings or in the street with the more expensive houses – he would have been too visible there for his liking.

When I walked back to the station, I saw that three dogs had taken up residence on the small patches of grass near the pine tree opposite the station building. A light brown dog kept watch, the other two slept. The dogs were familiar, like the other dogs I saw that day. I felt a desire to stay and live with them, similar to the desire I feel to follow the geese who migrate in autumn.

Olli had died three weeks before I went to Pașcani. He was fifteen years old and had been ill for a long time. In the last months of his life, my body was an extension of his. I helped him to get up and lie down, and because he could not hear and see well, we mainly communicated with touch. In the last weeks he only ate out of my hands. Olli was not sad about growing old. This was just the way it was, and his job was to keep going. I did feel the weight of it, but at the same time helping him was the only thing that mattered, and having something matter so much is a gift.

In this chapter I discuss the ongoing dialogue that Olli and I had, which is closely connected to our dialogues with our other companions, the dogs Pika and Doris, and the cat Putih. I wrote about life with Olli before, in 'Stray Philosophy' (2014), which focuses on the first months of our relationship, in which we established a common language and he taught me about forming new relations with others. While this was an important time, because we learned to understand one another well enough to live together and it changed my outlook on living with dogs, in the years that followed our conversation deepened and I continued to learn from Olli. In our dialogue touch played an important role, as did sitting together, and in what follows I will pay specific attention to these embodied aspects of the language-games we shared. In addition to the dialogue with me, Olli played a role in shaping the larger community of which we are part. In the second part of the chapter, I use these acts of his as a basis for conceptualizing multispecies political virtues, in order to investigate how other animals can take the lead in developing new forms of democratic government. One of the main lessons that Olli taught me was that dogs are perfectly capable of taking the lead in refiguring social and political life.

The role of touch in the conversation between Olli and me

Half a year before my visit to Paşcani, I cried in the car when we returned from the vet. Olli was feeling unwell – he had spondylosis and the pain medication did not always work well enough, and he also had a bladder infection that might be a sign of something worse. I sat behind the wheel, and Olli noticed my sadness. He sat up in the back of the car, which was not easy for him, and gently touched my shoulder with his nose. He always did that when I was sad or upset – he would come up to me, and touch the palm of my hand with his nose, or sit next to me and touch my shoulder. The touch was always very gentle and soft, sometimes I could barely feel it. This gesture says: I am here with you.

For Olli, touch was a very important ingredient of our communication. Touch is often overlooked in the study of animal communication and not considered to be part of 'language' (Botero 2017; Monso and Wrage 2021), but it plays an important role in the languages of many nonhuman animals (Meijer 2019). Philosopher Maria Botero (2017) writes that human researchers tend to focus on vision, but when studying the social cognition of nonhuman primates, touch should be at the center of attention. Through touch, primates learn to regulate their emotions and pay attention to others. Touch is important in education and communication, and essential in maintaining social relations. Monso and Wrage (2021) add that touch is important in the interactions between animals other than primates too, and that touch is not only relevant for the study of social cognition, because it is also often used to express moral emotions, such as love, grief, sympathy, and guilt.

In the conversation between Olli and me different senses played a role. He liked words and was especially fond of his name: the first months he was in awe of having a name, and someone who called him, which made calling him more important for me too. He also paid close attention to me visually, especially around dinner time. I used my eyes and ears to read how he felt, for example to know what he wanted, or to anticipate fear. Scent was important in his life, but not really between us – more so in relation to food he found on the streets, and other dogs. Taste mattered only indirectly between us, through food and treats. Touch initially mattered to him more than to me, as a means of communication, but during our time together it became more important for me too.

When Olli had just arrived from Romania, he was nervous. Stroking him in a specific spot between his ears was the only thing that always calmed him down. I learned this because he would sit down next to me on the couch, straight up, and invite me to touch him like this, by bringing his neck into the right position under my hand. After he taught me how I should touch him when he was nervous, he didn't need to ask me for it anymore. There were many other moments in which he came up to me for physical contact, in the park, the city, and the house. He often invited me to give him a belly rub by lying on his back in his dog bed or on the couch. He liked to lean on my leg and often offered me his paw when he sat next to me on the couch.

This degree of physical contact was new for me. Pika, my other dog companion at the time, and I were in full contact. We did not need many words, and there was a lot of casual touching between us – we would sit on the touch together, our sides touching, she slept on the bed and when it was cold our bodies would touch, when we travelled together on the train she curled up next to me – but she never really asked me to touch her. With Putih the cat, it was the same. He always slept on my lap when I wrote, and often around my head on the pillow when I slept. He was ill for a long time before he died, and touching him helped him calm down if he was unwell. But with Olli touch was much more explicit, especially in the first years. Asking me to touch him was his way of asking if things were still alright, and telling me how he felt.

When he was old, his sight and hearing deteriorated. He still saw movements and big objects, especially in daylight, and he could still hear some sounds, but much disappeared in the fog. Now it was me who used touch more explicitly in our relationship. I touched his side to direct him if he needed to go somewhere, or his shoulder to let him know I was near, or his face to get his attention around dinner time. I still spoke to him, but with my mouth close to his ear, so he could feel the words. For the spondylosis he received physiotherapy and acupuncture, and the physiotherapist taught me how to give him massages. He often asked for these, and touching him in that way helped me read his body and understand how he felt – if he was in pain, and if so, where it hurt.

Touching each other made our relationship safer for Olli, and it made our connection stronger. A touch was often the beginning of a sequence of events – Olli touched me or asked me to touch him, which could lead to play, cuddling, or going somewhere. My response to his questions was usually immediate, my body answered his without thinking. When I sat down on the couch next to him, my hand always first touched his fur.

Of course, being close to someone does not require touching. One practice I have come to appreciate very much throughout the years is sitting together quietly. Pika always lay next to me on the couch when I drank my early morning coffee and read. Putih was often on the other side of me. After Pika died, Olli took over this habit, and sat or lay next to me on the couch. The guinea pigs who currently live here, Simba, Klontje and Kruidje, do not appreciate being picked up, but they enjoy it when I sit with them. Spending time together in this way can lead to great attunement to others. Olli and I always sighed together – I mean the kind of deep sigh that expresses relaxation. I do not know if these sighs began in me or him. I also feel this attunement, that follows from being close to one another, when I walk with the dogs. I do not need to think or look in order to know where my walking partners are, I feel their presence near me, and adjust my pace to theirs. We all adjust to the terrain, to the weather, and while we sometimes make eye contact or use our voices, we mostly move as one without paying overt attention to others (Aaltola 2013; Smuts 2001).

Multispecies attunement, be it through touch, or moving together, or sharing spaces, requires time and attention. For Olli and me, touch has been central in gaining clarity about certain issues, and touching each other was at the basis of different language-games, including habits.[2] Developing

2 In *When Animals Speak* (2019) I discuss multispecies language-games and habits in detail in Chapters 2 and 3, as well as in Case Study 1.

new language-games together involving touch, often on Olli's initiative, who took touch more seriously than I do, enlarged our common lifeworld and continued to do so over the years (see also Bannon 2009; Haraway 2003; Merleau-Ponty 2003; Weiss 2006). Sometimes the full meaning of certain interactions only becomes clear much later, as with comforting him by touching his neck – I simply did what he asked and later understood how it helped him calm down. At other times it was clear immediately, for example when he asked for belly rubs. When he was old, he could not turn on his back anymore, but still liked it when I stroked the space between his front legs – he asked for that by moving his front leg in a certain way, barely noticeable.

Touch in multispecies dialogues

Taking touch seriously in multispecies language-games can help us understand how language is embodied, and bring to light nuances and complexities in common language-games that remain hidden if we only focus on sound and sight. It also directs us towards our way of being in the world more generally, as embodied, feeling creatures. Touch is not only underexplored in multispecies philosophy, but also in philosophy more generally. Two exceptions – the role of touch in phenomenology, and Luce Irigaray's ideas about touch as grounding life and expressing difference (2011) – make clear that there is much to be gained from taking touch seriously in multispecies relations, not only to improve these relations but also to better understand how we as living beings form understanding of others and of ourselves.

Different phenomenologists, notably Husserl (1952) and Merleau-Ponty (1962, 1968), emphasize the importance of touch in analyzing our experience because touch can help us understand the relation between self and world. In *Ideas II*, Husserl writes that touch is the sense that locates us in our bodies, and is reflexive – we can touch ourselves touching, for example when you touch your hand with your other hand. Touch enables us to experience our bodies as *Leib*, lived body, instead of just as *Körper*, the body as a mere material object. When you touch your body, you can feel your body being touched, and feel your body. This 'double sensation' makes you perceive your body simultaneously from the outside and the inside, as subject and object. According to Husserl, this grounds our self-awareness, our understanding of ourselves as bodies in the world, and provides us with a perspective from which we can think. The tactile body is part of the world, and it is public. Merleau-Ponty also points to this 'double sensation' and sees it as fundamental for our experience (1964, 168) (even though he later (1968) describes vision as a reflexive sense too). He draws attention to the fact that its reversibility (which means that we are both doer and receiver of touch) also happens when we touch another body, for example in a handshake (1968, 187, 142). Both bodies engaged in touch are part of the same intercorporeality, which is a bodily intersubjectivity. In other words: in touch we meet and connect as bodies, and this grounds the kind of beings that we are, our relation to the world around us and ourself.

Both Husserl (Dufourcq 2014; Painter 2007; Venuta 2023; Vergani 2021) and Merleau-Ponty (Dufourcq 2014; Toadvine 2007; Westling 2013) were concerned with the relation between 'human' and 'animal', phenomenologically and ontologically. I will not go into their views in detail here, because this falls outside of the scope of my investigation. What interests me is that Husserl and Merleau-Ponty highlight touch as a phenomenon that unites sameness and otherness both in relation to oneself – my hand touching my hand – and to others – Olli touching my hand with his paw. Touch spatially and materially grounds our understanding of ourselves and world, and it does so for humans and other animals in a multispecies world.

Olli actively used touch to position himself, which affected our interaction and his position in social life more generally: in relation to other humans and dogs, in homes and public spaces. He was very aware of his posture and bodily movements in relation to other animals too. His experiences on the streets, in different shelters and in a house with me and other companions, had taught him to be careful in social interactions, and he was very aware of the signals he sent out to others with gestures and his body posture. Touch helped him to figure out where he stood. At the same time, it was an expression of belonging, and at times, touch nearly made us into one body or being (Aaltola 2013), for example when we sat on the couch with our sides touching, as we often did. Touch affirmed our differences and played a role in the power dynamic between us, but it also connected us, and reminded us of our connection and of how similar we were.

The relation between touch and difference has been explored by Luce Irigaray (2011), who in this context makes a distinction between mental knowledge, which is what we usually mean when we speak about knowledge, and carnal knowledge. Through carnal knowledge we can learn about the otherness of others, and in this process touch is an act of individuation. According to Irigaray, touch does not adhere to the logic of possession that is often found in the act of looking. The act of looking follows the master-slave dialectic, she writes, because of the hierarchy between the spectator and the one at whom she looks, and because it is aimed at submission. In contrast, touch is connected to respecting the otherness of others, the place where the other lives. Irigaray calls this a 'sharing in difference' (2011, 140).

According to Irigaray, the potential of touch has been overlooked in the philosophical tradition, even in phenomenology, because many philosophers – here she includes Merleau-Ponty – favor sight over touch (Vasseleu 1998). She connects this to a more general approach of humans to the living world, which does not cultivate life, learn about or establish coexistence among different beings, but is separated from the body and from nature. In our thought systems and societies, existence is replaced by a structure of names and representations, a dead world. Humans created an artificial culture through 'a logic of representation' (2011, 133) which runs parallel to the actual moving and living world, and the present. A focus on touch can help us counter this.

Certain aspects of Irigaray's analysis are problematic, for example when she draws a strict line between 'humans' and 'animals' (2011, 134-135; Neimanis 2017). Or they are unconvincing: the act of looking can, for example, also be understood as witnessing – think of animal activists in the Save Movement, who go to slaughterhouses to bear witness to the animals who are waiting for their death in trucks. Their act of looking is not aimed at domination, but rather at paying respect to the animals, and acknowledging their suffering because there are no possibilities for helping them. But her analysis of touch as a tool for becoming aware of our embodied existence in relation to others, and as a starting point for conceptualizing relations without domination (Acampora 2004) offers entry points for developing new forms of embodied multispecies communication, and learning. Writing from the perspective of animal sanctuary, pattrice jones (2019, 2023) refers to what Irigaray calls the movement or flow of life, which gets lost in modern narratives of representation, as 'eros' and also sees it as a counterweight to domination. Touch as a force and way of being with others can also be a part of the bigger project of aligning our bodies with the environment of which we are always part (Bennett 2020; Neimanis 2017) – when we walk, we touch the earth with our feet, we leave a mark with every step.

At the same time, a multispecies lens shows that it is problematic to prioritize touch over other senses, as Irigaray proposes, or to designate any sense as foundational for the experience. Beings of different species prioritize different senses due to their bodily capacities, and humans with different abilities than the norm might do so too. However, including touch in analyses of language can help paint a richer view of human and multispecies language-games. In analyzing touch, it is not enough only to consider its existential dimensions, as I did so far. Because of the large power imbalances between humans and other animals, this also requires mapping the structures of domination that shape interspecies relations in this regard.

Touch and domination

Touch can play a role in developing multispecies relations of care and trust, building common worlds, and relating to others in embodied ways, but there are also dangers involved, especially in situations structured by human dominance over other animals (which means most of our society). Humans often feel entitled to touch animals and touch them for their own pleasure, which can be seen as a form of abuse, even in situations that we usually see as benevolent, for example with animal companions or petting zoos (Wadiwel 2017). Touch is also used as a tool of oppression in industries in which animals are abused. Steve Cooke (2021) draws attention to the fact that in animal agriculture the trust of farmed animals is betrayed by certain forms of touch. Farmed animals are social animals who are harmed by the fact that their relationships, such as friendships and parent-child relations, are standardly disrupted by the industry that abuses them. This increases their physical, psychological and emotional suffering and that is bad for business. Animals who suffer might harm one another or themselves; suffering can make them hard to handle, and reduce the quality of the products of their exploitation. Therefore, farmers develop ways of calming them down, such as 'gentle touching' methods (Cooke 2011). These methods are also employed in laboratories that use animals in experimentation. Cooke writes that gentle touching can play a role in ethical relations between beings of different species, but shows that in the context of animal agriculture it is ethically problematic, because it betrays the trust of the animals in question.

In Olli's case, his specific reliance on touch and using it as a form of asking me if we are still cool also follows from earlier harms. He had been severely abused by humans, and therefore needed to ask me things much more explicitly than most of the other animals I lived with. This was not the only meaning of touch: it also was a way to communicate about daily life, and to get to know each other. Furthermore, the role of touch changed between us as he became more comfortable in his new life. After a few years, he did not use touch out of nervousness anymore, nor to please me. Touch became part of our common repertoire of communication, and while our communication still included strong requests from him, when he asked for touch or touched me, it was to connect and show his affection. Still, with other humans he often relied on touch to test the water, and when he became frailer due to old age, he became more careful again.

My other dog companion Doris, who came to live with Olli and me in 2016, is less fond of touch. While Olli usually asked me to touch him, Doris and I take turns. I sometimes touch her when she is close to me, and she enjoys that. Now that she is older, she sometimes invites me to touch her tummy. She also touches me to invite me to play, and likes to play-bite me. She appreciates it when I do yoga and often takes part, by imitating movements or lying down on top of me. For Doris, touch is reserved for friends, it is not a tool to negotiate relations with others. In fact, she strongly dislikes to be touched by strangers. It took my parents a year to win her trust, and the vet about four years. Like Olli, Doris is originally from Romania, but she was already adopted by multiple families in the Netherlands before she came to live with me, and likely suffered abuse there (Meijer 2020). However, because she is a dog, many people want to touch her, at bus stops, in the streets, in the park, or wherever else we are. In fact, because she is a dog, they often feel entitled to touch her. Even when I tell them to not to, some will still try it, and when she says no by growling this is perceived by humans as an insult (Meijer 2020). Because humans come too close uninvitedly, Doris is always on guard when we are outside.[3] Living with Doris helped me to see how touch is often imposed on other animals – I knew this, but walking with her showed me how many humans feel they have a right to touch the bodies of other animals.

Touch and consent

While many scholars in fields like critical animal studies and animal philosophy emphasize the importance of taking into account animal agency and subjectivity and forming new relations, the question of consent is currently underexplored in the literature (Meijer 2020). This follows from the fact that centuries of oppression have normalized the use of animal bodies for human benefit, but it is also connected to the still common view that it is difficult or even impossible to find out how other animals really feel, what they want, and what they say. Learning to understand yes and no, and developing mutual language-games which could involve touch, based on consent, asks for moving beyond a position that I will call 'species skepticism'.

3 In the essay 'Throwing like a girl' (2005), Iris Young argues that humans of different genders carry their bodies differently in public spaces – as the title of the essay suggests, we can see this even in throwing a ball. Men use their whole body, Young writes, position their feet firmly on the ground, open their posture, while women throw only with their arm. While gender relations changed in the years that followed, her main point still stands: people who are not men are generally less safe in public spaces, and taught themselves to behave and move differently. Similarly, many nonhuman animals are unsafe, and both Doris and Olli developed ways of moving that anticipate human violence.

Species skepticism is a common epistemological position in western societies, which combines skeptical views about knowing others with speciesist ideas about other animals. Species skepticists assume we can know other humans, either because we are alike, or because we share a common language, but that we cannot know nonhuman animals because they are too different from humans and do not speak in human language. Their minds are like black boxes we cannot access. This position is problematic in different ways. First, there is no strict separation between humans and other animals: species differences are of degree and cognitively and emotionally there is much that humans hold in common with animals such as other mammals, birds, and fishes (e.g. Bekoff 2002, 2007; De Waal 2016). Second, knowledge of others does not simply hinge on species membership: social relations also matter (Donaldson and Kymlicka 2011). Through living with other animals, we can get to know them and they us (Calarco 2018; Haraway 2003; Howard 1952, 1956; Smuts 2001). Similar to how this works with humans, including ourselves, we cannot know everything about them but this is not necessary for understanding someone. In understanding, not only individual relations play a role, but also social and cultural processes (Crary and Gruen 2022; Despret 2016). Third, language is a bridge to others but can also deceive us, and speaking the same language is not a guarantee for understanding. Furthermore, even within languages there are many language-games, and we often do not share all of these with others. Philosophers and poets for example use language in very different ways, even when writing in the same language (Wittgenstein 2010, footnote 1). Fourth, presenting the mind as a black box obscures the role of the body in getting to know others and in language, as the earlier discussion of touch endorses (see also Acampora 2006; Aaltola 2013; jones 2019). Fifth, if we approach other animals as beings we can never get to know, we will never get to know them because we ask the wrong questions (Despret 2016).

Attending to actual dialogues shows that different individuals have different needs and desires with regard to touch. Doris and Olli do, and I also have preferences. Figuring out what other animals want requires asking them questions, allowing them to respond, paying attention to them, and responding to their questions. Here, consent works two ways: Olli also touched me, and sometimes asked me to touch him when I did not want to, so I said no. Both of us want the other to be well, so we like to say yes to each other, but sometimes he would ask me to give him a massage when I was working or very tired, and I refused by giving him a kiss on the head. He understood this well.

Further exploring the question of consent requires more freedom for other animals – like the possibility to say no and to leave a situation. This process begins with acknowledging that the question of consent matters and that humans do not have to right to impose touch upon others in interpersonal relations, as well as on the level of community, including in practices aimed at creating knowledge with and about animals.[4] While benevolent humans can work towards more equal relations with other animals in current societies, questions about consent and freedom also remind us of how deep the oppression of other animals reaches. One of the difficulties of being a human under the conditions of anthropocentrism is that when living with other animals, one is forced, in subtle and less subtle ways, to take part in their oppression, for example by holding them captive, using dog leashes and so on. Changing this not only requires activism to change laws and institutions, but it also asks for attending to how other animals are, and can be, agents of change. If humans develop new forms of multispecies co-existing without consulting other animals, they run the risk of reinforcing anthropocentrism, because this repeats epistemic oppression and a hierarchical view of relations between humans and other animals in which humans know best. Neglecting the perspectives of other animals in forming new communities is problematic from the standpoint of justice, and for epistemic reasons. The last part of this chapter therefore moves the focus from the individual relations between Olli and me to his position in society.

4 Blattner and Van Patter (2020) draw attention to the fact that existing frameworks developed to guide research relations with more-than-human animals, like the 3R's, are ethically inadequate. They write that many social scientists currently focus on research with nonhuman animals in households and sanctuaries, aiming to acknowledge their agency and subjectivitity, often with the animals' best interests in mind. But in order to develop this type of research in ethical ways, we need new guidelines. They offer three core principles: non-maleficence, beneficence, and voluntary participation. The last principle involves 'mediated informed consent and ongoing embodied assent'. While this may seem straightforward and logical, following this guideline is, in fact, far reaching, and includes ending research (and even relations) if animals say no. This does not mean that humans can never study nonhuman animals, or study the world with them. There are some examples in the literature of scientists who aim to study other animals on the basis of freedom and trust. I wrote about Len Howard's life with birds in this context, and the work of Barbara Smuts (2001) can also be seen as based on these principles. These scholars emphasize the importance of learning the other animals' languages in order to even know how they might say yes and no, and that attending to animals involves being open to changing yourself (see also Calarco (2018) on ethology and transformation, and the next chapter).

Dog political virtues as a starting point for forming new multispecies political practices

Olli was a pacifist. Peaceful relations with all animals (humans, dogs, mice, chickens, slugs) mattered to him and he was good at working towards these. Being polite and kind to others, avoiding conflict while standing his ground, and making connections, for him followed from having the right attitude. Olli did not calculate the greatest good for all, as in utilitarianism, nor did he follow a set of absolute rules, as in deontological approaches to ethics. Rather, his attitude can be best captured with a virtue ethical framework. He acted in response to the specific situations he was in, and learned to act in the right way by acting, or responding, in the right way – Aristotle refers to this as perfecting one's character. When we met, it soon became clear to me that he had his own norms and values, and that these had as much worth as mine. He was not acting on instinct, or only focused on his own gains: he judged situations, his acts were informed by his experience, and the unicity of the circumstances.

In thinking about ethics and politics, humans usually use human-centered notions of ethics and politics as the blueprint for 'ethics' and 'politics' (Probyn-Rapsey 2018). This influences not only our ideas about who is capable of ethical and political action, but also the content of these concepts. Anthropocentrism in ethical and political systems usually leads to the exclusion of other animals, and it obscures nonhuman animal forms or systems of ethics and politics as well as multispecies normativity. Nonhuman animals have their own expressions of normativity (Andrews 2020; Bekoff and Pierce 2009; De Waal 2016; Peterson 2012), as the brief discussion of touch as a moral sense in the beginning of this chapter also showed, and these can function as the basis of new relations with them.

In multispecies environments involving humans, like cities, towns, and rural areas, there is much translation and diplomacy going on between different communities: between humans and animal companions, between domesticated, liminal and wild animals, and between different communities of wild and liminal animals. When problems arise, or decisions need to be made, humans often background (Plumwood 2001) animal agency, and tend to want to make all decisions themselves. But other animals have their own perspectives on situations of common concern, and ways of navigating social relations.

To further explore this, we need more research into animal political projects and processes, as well as attention for how nonhuman animal communities and democracies are shaped and function. New fields of study, like political ethology and political biology, can play a role here. But we should also attend to how animals already co-shape existing multispecies communities and relations. Olli showed me the importance of political virtues in and between communities. A focus on virtues is helpful for rethinking multispecies political relations, because it enables us to recognize the embodied and situated character of political action, and because many nonhuman animals already learn and engage with social norms and moral decision-making in the way that virtue ethicists describe. Focusing on multispecies political virtues offers a new way of thinking about political agency.

Openness and kindness as political virtues

Two of the virtues that played a key role in Olli's diplomacy are openness and steadfastness. The virtue of being open to others, which implies attention to the other and vulnerability, is neglected in current political human landscapes (Latour 2018; Young 2002). In neoliberal capitalist societies, politics is generally presented as a struggle, and political engagement as a competition between predetermined interests (Meijer 2023a; Young 2002). This way of practicing and viewing politics threatens space for plurality and difference, which is foundational for the functioning of democracies. A full analysis of this phenomenon lies outside of the scope of this chapter, as does an analysis of countering it, which asks for a revision of how we view 'politics' (Meijer 2022c), including a revision of political language, and developing a different attitude towards others. Here I only want to show that attention for the practices of other animals is part of that process.

Attending to animal kindness and openness requires deconstructing 'survival of the fittest' narratives, which still inform ideas about animals (as 'wild' and acting on instinct instead of social norms), humans (and human nature), and nature/culture dichotomies. There are many counternarratives on which we can draw, for example in anarchist theory. In the first chapters of *Mutual Aid: A Factor of Evolution* (2021), the anarchist thinker Peter Kropotkin famously writes that many animals help one another, and that not conflict but cooperation is the basis of social life. jones builds on these views to argue for a 'natural anarchism' that connects humans and other animals (2023). Similarly, feminist philosophers of science and biologists have argued that a focus on struggle and conflict in biology and ethology followed not from neutral observations of animals but rather was the result of the respective worldviews of the researchers (e.g. Haraway 1991). In her conceptualization of other animals as fellow creatures, philosopher Mary Midgley (1984) also emphasized the relevance of cooperation in relations between nonhuman animals, and the continuities between humans and other animals in this regard.

For Olli, kindness and openness were always at the basis of encounters with others, not in an instinctual way, but as a learned attitude, in the way a human diplomat would use these qualities, and because he was a nice person who wanted to do right. I do not know precisely how Olli acquired his skills, nor how these skills played a role in the dog communities of which he was part before he came here. But in the Netherlands, his virtues played a role in our multispecies household, of which mice and guinea pigs are also part, in relations with my human family, in the larger multispecies community of the town where we live, and our former neighborhood in Amsterdam. Olli approached everyone in the same way. He greeted whomever he encountered, showing he meant well with his body language, while carefully observing the person opposite him. On our walks, we often encountered dogs about whom their human said: he is not good with other dogs. Or: he is not good with other males. But Olli never had problems with them. If he felt that the other dog was uncomfortable, he turned his head away, and made clear that he meant to avoid conflict with his body language. When we lived in Amsterdam, many children in our neighborhood were scared of dogs. He approached them in his friendly and careful way too, and sometimes I explained some of his body language, such as what it meant when he was wagging his tail. This led to friendships, and the children in the neighborhood learned to greet him as a friend. Because Olli did not discriminate between different types of humans or dogs, unexpected encounters took place. He gave certain humans that I would avoid the benefit of the doubt, which gave them the opportunity to respond in a friendly manner. While I still prefer a certain amount of distance to most humans, walking with Olli did teach me another way of relating to others. His protocol for dealing with others makes him a suitable candidate for negotiations and discussions of boundaries. After he overcame the fear that followed from his time in shelters, he also genuinely enjoyed meeting new people, human and nonhuman.

Steadfastness and the insights of street dogs

Olli's attitude of openness was combined with steadfastness, which combines strength of character with self-confidence. He had a calm presence because he had experienced many types of situations, and had made his own choices for a long time. Because he was used to making his own decisions, he never looked at me for assistance or judgement – perhaps he discussed matters with the other dogs he lived with, but not humans. He never engaged in activities that did not interest him, let alone be lured into joining others in violent activities. When dogs bullied another dog, as groups of dogs sometimes do in parks, he never participated and, when possible, tried to intervene, and when humans asked him to participate in activities he was not interested in, like playing with a toy or ball, he would sometimes give it a go to show his good will, and then ignore further invitations.

In the Netherlands, making one's own choices is an attitude which is not generally appreciated in dogs. Even though training practices are changing, from models of domination and obedience to partnerships (Wlodarczyk 2017), companion dogs are still subject to disciplining practices, ranging from laws that restrict their freedom of movement, to social expectations of obedience (Meijer 2014, 2020). Processes of domestication also influence the kind of agency dogs have, leading to large differences between, for example, labradoodles and Romanian dogs.

There are not many dogs in shelters in the Netherlands (NOS 2017), and breeding is increasingly criticized in politics and public discourse because of congenital defects in different breeds (NOS 2017). This has led to a rise in the adoption of dogs from Southern and Eastern Europe (NOS 2017), who often need time to adjust to their new situation and generally are what Sue Donaldson and Will Kymlicka (2011, Chapter 7) call 'liminal' instead of domesticated: they live amongst humans and may depend on them, but are not used to the close relations that are expected from house dogs. Recent newspaper coverage in the Netherlands focuses on problems with 'foreign dogs' (NOS 2021), that include specific diseases that these dogs may have, such as Leishmania, but also behavioral problems. These problems may involve fear or anger, and mostly follow from the transition that the dogs are forced to make: instead of being thankful and happy, as humans expect because they think they saved the dogs, they resist oppression, are not automatically obedient, and generally want to keep making their own decisions.

Instead of viewing these dogs as problems, we should appreciate their perspective on the systems of domination that we are so used to that we do not notice them anymore. Because these dogs are their own people, they challenge hierarchies, restrictions, and human conventions. Our societies are made to fit humans, which comes at the cost of animal habitats and peaceful multispecies cohabitation. But this attitude is biting humans in the tail; the challenge for humans in our age is to refigure what it means to be human, and this involves taking seriously the life-worlds of others with whom we share this planet (Celermajer et al. 2020, 2022; Tschakert et al. 2021). The perspectives of liminal dogs on existing structures can inform advocacy and point us towards change (Lemon 2015).

Here, it is important to realize that the fact that these dogs have other normative systems does not mean that they do not have norms at all. Olli was friendly, polite and gentle, and not wild or aggressive. But he definitely was his own person, someone who grew up in freedom, and who only in the last months of his life came to rely on me – which was a beautiful thing too.

While openness relates to the values of cooperation and kindness, in Kropotkin and Midgley's sense, steadfastness is connected to holding one's ground and critique. These are not the only virtues that can play a role in forming new multispecies societies. Different dogs have different preferences with regard to social and political interaction – similar to humans. Mapping these can be part of further exploring how embodied virtues play a role in and between communities, and the jobs that nonhumans find for themselves, which should be part of rethinking democracy in a multispecies world. A focus on animal virtues, for example by exploring the practices in which they are embedded, can function as a starting point for rethinking relations from the ground up.

Greeting

One practice in which virtues are expressed is greeting. Greeting is an important practice for many animals (Meijer 2013, 2019; Smuts 2001), as well as a ritualized political practice in different human cultures (Meijer 2013, 2019; Smuts 2001; Young 2002). When you greet someone, you acknowledge them as 'someone', and therefore greeting practices can open up the way for further discussion. They can play a role in conflicts, because both parties pronounce their good will, and can be a step in better understanding the position of others – which does not automatically lead to harmony, it can also be part of resistance, or ongoing arguments.

Olli always greeted everyone, as a matter of respect and politeness: humans, dogs, cats, chickens, and small animals, like mice or bees. For Olli, greeting had multiple functions. He showed he meant well and investigated how the person opposite him felt about interacting. Depending on how the other responded, greeting became the opening of a dialogue, or the end of the encounter. Dogs usually responded to greetings, but humans sometimes kept ignoring him, either on purpose or because they did not notice him, or even made fun of him. This did not bother him, and he would again show that he meant well in the next encounter, though his enthusiasm would wane. If we encountered humans who were scared of dogs, he would approach them more carefully, or even ignore them. When he had lost most of his sight, he could not pick up on their signals anymore, so I had to tell them that he was old and meant well. Only at the very end of his life did he begin to avoid greetings with strangers, because he could not read their bodies anymore.

Developing better greeting practices can play a role both in the transition to more just multispecies communities and in these new communities (see also Meijer 2013). Greeting is reciprocal, and based on the willingness and acts of different actors. Creating new collective habits, such as greeting, is an easy way of building better multispecies relations in which animals of different species can show their attitude towards engaging with one another. There are already many individual greeting practices to build on (see for example Cornips 2022 about cow greetings). Groups of street dogs also have communal ways of greeting humans, in which individuals choose specific roles. Patterns of greeting may change and improve over time, as different parties feel more seen, understood, or simply safe. Greeting processes can help establish boundaries between communities, but also function as a starting point for discussing other, perhaps difficult matters.

An ongoing conversation

In the nearly ten years that Olli and I were together, we spoke about many things – walks, food, comfort, joy, our hopes, even death. He was with me when our companion Pika died, and Putih, and my father too, all of them were his good friends. He comforted me when I was upset, and I him – he had a great fear of thunder and fireworks, and there were other fears connected to his past that never fully left him. He experienced many difficulties, including his ailments; life never became perfect for him. But it was a full life – not many dogs get to lead two lives in one – and in the last year he felt very safe. He loved his dog bed, and the garden, and our walk, and the car. The car was the summum of luxury for him: to get somewhere without walking, while being completely safe and in close proximity to Doris and me.

Olli had a good death. Our vet, M., was a friend of his. We went to see her the day after he arrived in Amsterdam and she treated him with love, which he never forgot. She also had a special place in her heart for him. When Olli told me it was time, I called her to make an appointment for the next morning. That night I slept on the floor downstairs next to his dog bed, with my hand between his front legs. This was the most beautiful night of my life. In the morning we sat together. When M. came, Olli felt honored – after all these years of visiting her, she finally came to visit us. She gave him many treats, he went into the garden to pee, it was all really joyful, and then he climbed onto the couch and it was time.

Through the ongoing dialogue between Olli and me, my view on life changed. I changed – my priorities shifted through learning to appreciate what he valued, like sitting together, making little jokes, enjoying the garden in all seasons, meeting friends, returning home together. He could follow life well, align himself with what it asked from him. And he showed me who I am through his eyes. Many of my memories of Olli are very moving, mostly because he was always trying to be a good person for others, even if they did not appreciate him or see him for who he was. For a long time, the worst thing I could think of was that I would never have found him, that he would have spent his life in the shelter, or worse. Now that he is not with me anymore, I can see the full story of his life, at the least the part of it he spent with me. And his commitment to life, to going on regardless of what happens. When somebody close to you dies you become a stranger in your own life, not unsimilar to the stranger I was in his town and he in mine. But I know what he would say to this, which is to simply keep going, because that is what life asks from us.

CONVERSATIONS WITH MICE

I play a tune on the ukulele and sing. The ex-laboratory mice who live in the room upstairs begin to move, they like this tune. They run up the stairs of their home, and back again. The next tune I play is gentle, they calm down.

'Hello mice,' I call into the room.

Some of the mice are awake and look at me intently. Vachtje pokes her head out of the sleeping box.

I give them rice, they really like rice, and now everyone comes out of the sleeping box to eat.

'Hello mice,' I call into the room. I was away for a few days and my friend G. took care of the mice in my absence.

All of the mice come out of their sleeping box to greet me.

One night as I play the piano, I see a small face behind the books on the lowest bookshelf. It is the house mouse, who from then on comes out to watch when I play. Maybe they feel comfortable to do so because I am harmless when I play, or because the lab mice listen and it changes the atmosphere of the room. If I move away from the piano, I become a human again and the mouse leaves.

Spokie is very old. Her companions died, and even though she was not very sociable – she preferred to sleep away from the others, in the tunnels – she did go looking for them after they died. Whenever I see that she is awake I go up to her to speak with her, to keep her company. I call her name, sing songs, and she responds by looking at me and climbing up to the first floor. At first, she does not like being touched, but when she is alone for longer, she begins to ask me to very softly stroke her ears, by putting the side of her face to the ground. I recognize the gesture: this is how mice ask others to groom them. It is difficult to stroke her in the right way, because my touch needs to be very light, but I learn how to do it.

3. Learning to See Mice[1]

Abstract: I discuss here the conversations I had with a group of ten female ex-laboratory mice who came to live with me in the summer of 2020. These mice taught me that mice are not the kind of beings that humans think they are, and about what matters in life more generally. In our conversations, sound and music played an important role, as well as material interventions and co-creating new habits. In the chapter I not only describe how the conversations between the mice and me developed, but also discuss the politics of living with mice, drawing on insights from (vegan) ecofeminism, and possibilities for mouse-human deliberation, drawing on multispecies deliberative theory.

Keywords: mice politics, mice dialogues, mice deliberation, animal testing, ecofeminism, laboratory mice

This chapter is dedicated to Breedsnuit, Bullie, Flankie, Grote muis, Kleine muis, Kleinoor, Kraaloog, Mooitje, Vachtje and Witoog, and to their successors Sneeuwtje, Stompie, Wolkje, Vlokkie, Madelief, Tweeteen, Mus, Maanoor, Spokie, Lieve, Roodmuis, Zeetje, Neushoorntje, Bram and Wezel.

When Vachtje (Furry) could not run in the running wheel anymore, she decided to sit next to it and run with her hands. The ex-laboratory mice who live in my house do this if they want to join someone else in the wheel, in order to judge the speed with which the other is running, to get a feeling for it before they jump in. Vachtje had a tumor in her left hind leg, making it impossible for her to run in the wheel, though she could still walk. By running with her hands, she could still take part in the activity. Vachtje's hand running shows that mice like to run in the wheels, but there is more to

1 This chapter is based on the article 'Learning to See Mice: (Stray Philosophy III)'. *Humanimalia, 13*(1), 203-251.

Meijer, Eva. *Multispecies Dialogues: Doing Philosophy with Animals, Children, the Sea and Others.* Taylor & Francis Group, 2025
DOI: 10.5117/9789048564415_CH03

it. Running in the wheels is also a habit, and something that gives meaning to their life.

The ten brown female mice who live in the mouse house upstairs were born in a laboratory at the University of Utrecht, Netherlands. They were bred to be used in experiments, but never took part in any. Usually, such mice are killed. In the Netherlands 159,614 mice were used in experiments and killed in 2019, and in the same year at least 262,238 mice were killed who had been used for breeding or who had been bred but not used in experiments (NVWA 2020).[2] Together with around a hundred other rats and mice, these mice were part of a pilot project in which laboratory mice and rats could be adopted by the general public.[3] This project is aimed not only at helping individuals, but also at changing public opinion about laboratory animals by drawing attention to their subjectivity.

The mice were born in February 2020 and came to live with me in August of that year. In the time that followed, I watched them often, and interacted with them on their terms. In the beginning, they did not like me much. They thought my hands smelled bad, but they did like my voice and came out of their sleeping houses when I called them. Because they liked my voice, I began playing songs for them on the guitar, ukulele, and piano. I do not always play for them; sometimes I just sit and watch them. When I wrote this chapter in January 2022, three of the ten mice were still alive: Kleinoor (Small Ear), Witoog (White Eye) and Bullie (Little Bull). They walked around freely in the room, sometimes using my body as an object to climb on. Witoog also liked to sit by my side.[4]

Through engaging with the mice, my ideas about what kind of beings mice are changed completely. I was struck by their practices of care, their ways of interacting with each other and with me, and I learned much from them, about mice, but also about what matters in life. I began to search for studies about their social lives, but found none. Laboratory mice are generally studied in order to learn more about questions that concern humans, and are seen and treated as replaceable. In this chapter I want to challenge this view and bring the actual animals, their personalities and social relationships, into view. In particular, I focus on their individuality, sense of community,

2 For empirical, moral and political problems with animal testing, as well as alternatives, see Herrmann and Jayne (2019).

3 This project was a collaborative effort of the following Dutch animal rights and animal welfare organizations: *Dierenbescherming, Animal Rights, Knaagdierenopvang Het Knagertje, Stichting Hulp en Herplaatsing Huisdieren, Instantie voor Dierenwelzijn Utrecht.*

4 Kleinoor, Witoog and Bullie have now died, and I adopted two other groups of mice since I wrote this text, who have also passed away.

practices of care, and ways of creating meaning. I do so by alternating narrative – telling stories about the mice – and philosophical reflection. This mirrors my attitude towards the mice: I look at them and what they tell me, then reflect, then look again.

In the first part of the text, I give an impression of what I saw when watching these mice. I then contrast this with existing views of mice, in the laboratory and in culture more generally. In the final sections, I propose to view the interactions I have with the mice as dialogues that can play a role in changing the political circumstances that determine the lives of mice. My aim in this chapter is quite modest, however: I want to give an impression of the mice that I was fortunate enough to get to know and explore new ways of thinking about mice-human relations, not to provide a full account of what it means to be a mouse, or to live with mice justly. For that, society would need to change. This text can only offer the first steps in thinking and living differently.

Life in the mouse house

The mice live in a house with two floors that contains several running wheels, one or two large cardboard boxes for sleeping, and other smaller boxes for hanging out, hiding, eating, or climbing on. They have different stairs for getting to the top floor, and sometimes I give them a so-called 'snack labyrinth', which takes up most of the space on the ground floor. There are always many toilet rolls to walk through or hide in. I also make objects our of clay for them and give them willow branches to climb and chew on.

The mice are awake around dusk and dawn. Most of them sleep during the day and during the early hours after midnight, but sometimes someone wakes up in daytime and runs in a wheel or eats something. When only one or two mice are awake, that is all they do; when most or all of them are awake they have social encounters, play, and are more eager to explore and try out new activities. Mice are always working on projects. Besides eating and sleeping, their main activities are running in the wheels, working on their nests by collecting scraps of paper, nest materials, and hay, exploring or inspecting the house (spaces often need to be inspected again), and social events like grooming.

I scatter their food twice a day on the floor and in the different boxes, and after that they spend time looking for the foods they like and eating. Individual mice have specific preferences. For example, some like zucchini, others not; about half of them like chickpeas; most are fond of dandelion leaves; everyone likes breadcrumbs, rice, pasta, oatmeal, and nuts; most like bananas; nobody in this group likes apples, but I know other mice who do; they do not really like carrots but will eat them. Some mice like many foods, like Bullie and Vachtje, while others are really picky, like Mooitje (Little Beauty) and Kraaloog (Beady Eye). They also have different habits concerning how much they eat and how much time they spend on it. Kraaloog and Kleine Muis (Little Mouse) prefer running to eating. They eat a bit when I give them the food but soon continue their activities and will eat a little something later. Others spend hours looking for the exact right nut or grain. Bullie just eats a lot.

The same applies to running. The first time I put a running wheel in the house they immediately liked it, and I understood I had to get more because they all wanted to run at the same time. They prefer the wooden wheels to plastic ones. I had to throw the wooden wheels away because of a red mite infestation, and bought plastic ones instead. But in the weeks that followed they did not run much and gained weight. I bought wooden wheels again and the mice ran the whole day; this was clearly a joyful event.

As Vachtje already showed, when the mice want to join someone else in the wheel they first run with their hands, and then just climb in, which often disrupts the other's running. They might try to climb up in opposite directions for a while until one of them leaves. After a year, they learned to run side by side in the bigger green wheel. When running together does not work out, sometimes the mice involved will start grooming one another. When a mouse falls out of the wheel because someone else is heavier or runs faster, they will wash their face for a couple of seconds to recuperate.

The mice are very specific about their boxes, using nesting material, cardboard, paper, and hay to build nests in them. Some materials need to be in some of the boxes, and other materials need to be removed. It is not always clear to me why some bits of paper should be in the nest and others not, and when I try my hand at it, I always get it completely wrong in their eyes. The mice have clear ideas about this, and never have a difference of opinion about it. I give the mice many small cardboard boxes to hide in and climb onto. After a day or two I understood that they want not one but multiple doors in their boxes, but it took me months to understand that they also really like holes in the ceiling (with the exception of their sleeping house). Because they are inquisitive, I add or remove objects daily. While most mice build their sleeping nests pragmatically, some really make an effort to create the right shape. Bram and Wezel (Weasle) in particular made beautiful flower shaped nests when they were old.

In addition to more serious activities like eating, nest-building, and running, the mice like to play. Sometimes one of the mice just runs all around the house. They also enjoy moving when I play certain songs on the ukulele; they run over the roofs of houses and boxes and create different routes through different boxes. I clean their house once a week and usually do not take them out when I do so, because they don't like it when I touch their bodies, so I first clean one floor and then the other. Afterwards the mice are really happy (unless I take away their sleeping nest: this upsets them and they immediately begin working on a new one). All of them are awake, everything is in a new spot, and they explore the whole new set-up together. Some run in the wheels, others climb over the houses, everyone goes up and down the stairs multiple times. This is a social event; the joy is shared. The ritual used to be quite exuberant when they were younger, now they are more thoughtful and slower. They still make jumps of joy, like foals or rabbits do.

The mice generally sleep together, sometimes with the whole group in one cardboard box and sometimes in smaller groups, though usually not more than two groups. When it is really warm, one or two mice might find a place alone on the top floor, but most will still join the others. They sometimes sleep on top of each other.

Some mice are more solitary than others. They might still join the others for sleeping or grooming, but also like to spend time alone. Witoog, for example, likes to be apart from the others sometimes, spending time upstairs when the rest of them are downstairs. Now that she is old, she sleeps a lot and does not avoid the others anymore, but she still likes to keep to herself. Mice who are ill may also become more solitary. For example, when Vachtje's tumor was big she was awake more and spent more time looking for food while the others slept. Kleine Muis and Kraaloog often run in the wheel when others sleep, but this has to do more with their temperament – they like to do a lot of running – than with the fact that they like to be alone, because they are often with the others and join the collective activities.

The mice never desired much contact with me. When I went to Het Knagertje to pick them up, I only had experience with liminal mice who lived in the wall of our former house. These mice had had some experience with humans before they came to live with me, and were not afraid of me, nor violent towards me. But they were also not eager to establish a relationship. Early on, I sometimes put my hand inside their home, but they regarded it as a foreign object to be buried with nesting material. They will still smell my finger when I hold it near them, but then they move away. However, they have individual preferences. Vachtje liked being touched from time to time.

Witoog liked me most in the beginning and still likes me best, although Kleinoor nowadays also does not mind when I touch her. Witoog also likes to sit next to me when Bullie and Kleinoor are exploring the room. Bullie still prefers not to be touched by me, but she does use my body for climbing on.

The mice communicate using touch, taste, sounds (they squeak, but only very rarely and softly in my hearing range, and make a sort of chewing sound, 'chuck, chuck'), sight, and scent. They also make use of tail and ear movements, facial expressions, gestures, and other movements, such as mimicking the movement of others. A few examples: when someone holds her tail up straight into the air this means she is excited. The mice use the chuck chuck sound when they explore. They often kiss each other on the mouth when they run into each other, which looks like a greeting. I will get back to facial expressions in more detail below, but the reason I knew they did not like my hands was the look of disgust on their faces. They like to sit beside someone else when eating, sometimes very closely, with their sides touching. They also like to sit next to others and do nothing, and sometimes they sit tail in tail. The mice do not like to step over someone else's tail, they will move a tail they come across with their hands so they can walk around it. Their movements are generally elegant.

Mouse language encompasses (ultrasonic) sounds and scents, but also gestures, movement, and grooming. The mice who live in my house have their own language-games, but also developed new ones in relation to me. An example of such a language-game is an argument. The mice argue occasionally. Their arguments usually last for one or two seconds and are very rare; in fact, I have witnessed only a few. Because their lives are short, they very rarely live longer than two years, everything happens fast: learning, mourning, arguing. The arguments were about food, though they usually do not mind when someone else steals their food and will just look for something else. Arguments include loud noises, gestures, and bodily movements, and are easy to understand for a human witness. Another example of a language-game is greeting. The mice have different ways of greeting one another – like kissing someone on the mouth, or aligning their movements with the other's. When a mouse meets another mouse, they might give up their own activity, and follow the other, mimicking her movement or posture. They also have ways of greeting me, acknowledging my presence by making eye contact or just coming out of their sleeping houses when they hear me – these are new, shared language-games.

Practices of care

The mice have many ways of caring for one another. Their most obvious practice of care is grooming. They spend a lot of time grooming others, in pairs or threes. When they were younger, they groomed one another in the cardboard boxes and sleeping houses or in the wheels, and later on they also did it in plain sight, but when they are roaming freely in the room, they use the space underneath my legs.

Grooming can be brief or take a long time (up to ten minutes) and is often reciprocated, around sixty per cent of the time. The practice ends when one of the two mice leaves because they want to do something else, or when a third mouse joins them and the one who was doing the grooming now turns to the new person. Grooming means washing someone else, the mice often using their hands to hold onto the other's skin. This can be for balance, or because the work requires it, for instance when you are washing someone's ear. Ears can take very long to clean. Grooming can be very gentle, or more like a massage. Some mice really relax when they are being groomed and may put their face flat to the floor (looking like humans receiving a massage). The mice are very clean and never smell bad. But grooming is not just practical, it is also a social practice that changed over time. When the mice had just arrived, they did not groom much, perhaps because they had had no houses, hiding spaces, or wheels in the laboratory and they did not feel safe enough for grooming, or did not have enough room for it. Later they groomed a lot, and this continued throughout their lives.

Mice also care for others who are ill. When Vachtje was ill, she once ran into Kleine Muis in the snack labyrinth. Kleine Muis put her front paws around Vachtje's neck, and gave her some kisses. This was a hug. She then climbed up the wall to make space for Vachtje, who could not climb anymore because of her tumor. I also saw Witoog give Breedsnuit (Broadsnout) a hug when the latter had a big tumor. She put her arms around her, gave her some kisses on the neck, and moved on. When Mooitje suddenly lost a lot of weight for reasons unknown to me, Vachtje often sat very close to her. Mooitje recovered in a few days because I fed her some foods she really likes, mostly rice. After Bullie had a stroke, Kleinoor went up to her often to give her kisses on her snout and on her neck. Witoog sat with her and crawled over her. This happened when only Bullie, Kleinoor and Witoog were still alive.

I have also twice seen four or five mice surround a mouse who was ill. They briefly formed a circle around her, a circle of support, and then went on with their business.

Bullie always sits with mice who are ill and does not leave their side when they are dying. I usually take her with me when someone needs to go to the vet. When a mouse is very ill and I know she does not have long to live, I keep an eye on Bullie to judge the situation.

Care continues after death. Vachtje, Breedsnuit and Kleine Muis all died in a single week, in July 2021. The others did not seem to respond much. When Mooitje and Grote Muis (Big Mouse) died on 1 September of that year, the others were scared and shy for a week. They did not eat well and hid in their boxes. Flankie was the sixth to die, and after her death the other mice tended to the body. Her friend Kraaloog often went up to her to greet her after she died, and finally pulled her into a corner by a leg. Kleinoor and Witoog groomed her and then buried her with the nesting material. In the two groups that I adopted after this group, I witnessed the same pattern. There is not much response to the first deaths, then they understand what happens, because the mice do not return, it scares them, and then the remaining mice develop practices of death care: they greet the body, then wash it, then bury it. I sometimes have to search for a mouse who died, because the others buried her so well.

Ecofeminist scholars (Adams and Donovan 2007) often draw attention to care as a fundamental axis for reconfiguring relations with other human and nonhuman animals and the natural world. A focus on care foregrounds relationality and interconnectedness. This provides an alternative to liberal modes of being that view human subjects as atomistic and autonomous agents, and systems in which humans are positioned hierarchically above other animals. Care can be both a theoretical lens and a praxis. An ethics of care encompasses situated and embodied attentiveness towards individuals, as well as attention for exploitative economic and political structures and a commitment to change these structures.

On the personal level, the mice and I are connected through daily relations of care, because I feed them and make sure their house is clean, and we communicate. As Josephine Donovan (2007) writes, relations of care with other animals should be dialogical and include their perspectives. When the mice are older or ill, the daily care intensifies: I pay more attention to the physical wellbeing of the mice, and they are more prone to illness, so they need more support from me. Flankie needed antibiotics and painkillers for a while, and mice with tumors may need more or different stairs in the house, and wider openings in the sleeping boxes. By keeping me company, the mice care for me too. This is not a symmetrical relation, but they are responsive towards me and acknowledge my presence. I also share some of the care with other humans, by telling them about the mice, and discussing the highs and lows with them, through which they care for me too.

Caring for the mice is not only nice or friendly. It involves holding them captive, giving them medication against their will, and in general controlling many of their life choices. They did not ask for my interference and at times show that they do not want it, for example by resisting when I try to give them their medication, or by trying to escape. Similar to how this works with my dog companions however, my agency here is also limited. I cannot end the use of laboratory animals singlehandedly, and we all live in a world hostile to mice and most other animals. My job is to remain attentive and support the mice's agency and wellbeing where I can.

For the mice, community is central, and care is an expression of community. Socially and emotionally, they care well for one another, and therefore do not need me. How this plays out differs from mouse to mouse. Bullie takes grooming extremely seriously. She used this to comfort other mice in their final moments, and by accompanying them to the vet and staying very close to them. Others, like Kraaloog, have a more casual way of being with others. But she often went to greet her best friend Flankie when Flankie was ill. She died not long after Flankie died. The others did not stay with Kraaloog when she died; she chose a space apart from them.

What is a mouse?

Mice are communicative beings who actively form meaningful relations with creatures from their own and other species. Through watching them and speaking with them in embodied and material ways, I learned about their social relations and individuality. What they show me stands in stark contrast to cultural views about mice.

Both cultural ideas and scientific knowledge about mice are in large part shaped by prejudice and human domination. According to Google, for example, a mouse is someone to be killed. When I search for 'mice', the first page gives twelve links to pest control companies, and two Wikipedia pages. ('Mouse' in Dutch gives me a page full of links to computer mice.) Only on the third page is there a link to the Dutch organization for responsible pet ownership. How humans see different groups of mice is generally based on their use for humans. Laboratory animals or companions are seen as useful, wild mice are generally respected or even admired, and liminal mice, such as the ones that take up residence in human houses or gardens, are considered to be pests.

This is not unique to mice. Knowledge about other animals often reflects their use for humans. Humans in most parts of the world currently live in societies that are anthropocentric. In this worldview, humans are not only the most important animals, they are also the standard by which other animals are measured. In certain areas – such as law or politics – this goes hand in hand with a strict distinction between humans on the one hand and all other animals on the other. This hierarchy is inherently violent: Dinesh Wadiwel (2015) calls this type of violence 'epistemic violence'. Epistemic violence serves to legitimate other forms of violence, because it makes them invisible. In the case of mice, epistemic violence legitimates institutional violence, such as experimenting on mice as if they were objects, but also direct violence – maiming and killing liminal mice is seen as completely acceptable by most citizens and institutions.

Current scientific research usually reinforces epistemic violence. Processes of knowledge production are not neutral, but intertwined with political, social, and economic structures. Neoliberal capitalism, for example, has an effect on the objects and methods of study, because it favors knowledge that is deemed economically useful. Anthropocentrism also affects knowledge production.[5] Scientists have long used other animals to gain knowledge about humans, or other topics relevant to humans, instead of trying to find out their perspective on the world or relations. Earthworms, for example, are often used to study soil and ecosystems, but not to find out anything about earthworms (Meijer 2019, Chapter 6). Similarly, mice are used to study human diseases, the origins of human emotions, and many other topics, but not to find out more about mice perspectives on a shared world. When their behaviors, emotions or cognition are studied, the outcomes are influenced by the material conditions under which knowledge is generated, such as isolation and captivity, electroshocks, and genetic manipulation.

The scientific apparatus affects mouse welfare and knowledge production about mice, as well as the agency of human researchers. Klaus Amann (1974) traces how mice and other living beings are transformed into 'technofacts' in

5 I should note that this way of researching animals is increasingly criticized. Ideas about how humans can get to know other animals are changing rapidly in different fields of study. In areas such as biology and ethology, there is more and more attention for the animal perspective. For example, in studying animal languages, human language is no longer taken as the blueprint for what language is, but instead species-specific languages and modes of expression are studied. Furthermore, researchers are becoming increasingly aware of their own position and of how the material, social, psychological and other conditions under which the animals in question are studied may influence the outcome of studies. In the humanities, a similar move can be detected. Instead of simply looking at the information we have about other animals – for example concerning their cultures, emotional lives, forms of cognition – in different branches of critical animal studies the political and social frameworks in which this information is generated is also taken into account, including power relations between humans and other animals. Still, the large majority of the studies in which mice and other animals are used are for human purposes, which shapes the type of knowledge that is created about mice and reinforces their position in society.

the laboratory. In DNA research, for example, the structure of the scientific machine translates real animals into a certain kind of information. This translation follows larger scientific-political structures. Tracing how specific forms of knowledge production form different images of mice, or produce different kinds of facts, can help to make visible power relations and show how agencies are intertwined. But getting a clearer view of, and responding to, the actual mice involved also requires attentiveness to larger political structures, as the study of mice's facial expressions and Donna Haraway's discussion of the OncoMouse shows.

Responding to mice: Facial expressions and the Oncomouse

Mice are the most commonly used animals in experimentation in the Netherlands. While they are used in studies aimed to benefit humans, this has also led to knowledge about their bodies and minds. For example, empathy and care for others have been studied extensively in mice. Studies show that they feel each other's pain and fear, and will console others when they are in pain (Pierce 2008). These studies often involve hurting mice and making them watch others being hurt, but the more benevolent studies also result in death – there are no large-scale rehoming programs for mice.[6] Andrews and Monso (2020) write about a similar moral problematic in the case of rats. Rats are used in experiments that prove their moral worth within the language-game of western science, but they are still killed afterwards.

The problems with these studies are not just moral, but epistemological too. Vinciane Despret (2016) shows how researchers' views about animals often determine the outcome of studies. Research questions often reflect prejudices related to identity categories, such as species and gender. Because research questions set the frame in which animals can answer, production of knowledge that is uncritical about power hierarches produces knowledge that tends to reaffirm the social status of the beings in question.

An example of how the way in which research is set up affirms stereotypical views of mice is the study of mice facial expressions. Researchers at the Max Planck Institute of Neurobiology in Munich, Germany, studied facial expressions in mice and were able to connect five emotional states to these expressions: pleasure, disgust, nausea, pain, and fear (Dolensek, Gehrlach, Klein, and Gogolla 2020). Using computer algorithms, they could also measure the relative strength of the emotions. The goal of the study was not to understand mouse emotions better, but to investigate the basic

6 This would make an interesting meta-study: what do studies about mice empathy show us about the empathy human researchers feel for others?

mechanisms of how emotions are generated and processed in the human brain. Mouse emotions and expressions are described as if the mice were machines. For example: 'Mouse facial expressions evoked by diverse stimuli could be classified into emotionlike categories' (2020, 89), or '[t]o study facial expressions, we exposed mice to a diverse set of sensory stimuli that can be assumed to trigger changes in emotion state' (2020, 89). Describing mice in this way influences the outcome of the study. If you study mice as machine-like beings, this will have an effect on what you see, or what the computer sees. It is also important to note that the researchers themselves were not able to recognize these mice's expressions; they did not take the time to learn this, so they used computer algorithms. The descriptions of the mice used in the experiment are based on prejudice, but also reinforce views of mice as being replaceable and acting more strongly on instinct than humans.

There are many problems with this study, such as the violence inflicted on the mice, the focus on human benefit, and the mechanistic, Cartesian conception of animals. But it also does not begin to do justice to mice facial expressions. For me, it definitely took time to get to know the mice. But some of their facial expressions were clear from the beginning – such as the look of disgust they gave me when they smelled my hand. Now that I know them better, there are many things I can read in their faces. I can see if they are curious, excited, hesitant, satisfied, uninterested, concentrated on food or nesting material, excited with big eyes after running, greeting others, or tired. Their sleepy faces, when they come out of the sleeping house after they have just woken up, are worth a special mention. Describing facial expressions in the way that the study does by no means begins to do justice to what they can tell with their faces, even to this human observer who misses much.

The Cartesian perspective on mice is challenged by Donna Haraway (2008, Chapter 3), who discusses the OncoMouse, a type of laboratory mouse that has been genetically modified to be a breast cancer model for humans. Haraway criticizes viewing these mice as a sacrifice for human reason, as well as perspectives that condemn breeding and killing OncoMice as a form of domination. Instead, she proposes to view laboratory practices, such as the ones in which OncoMice are created, used, and killed, as a historical and contingent inequality in which the multiplicity of the world is reflected. Humans and mice are entangled in multiple ways, and Haraway regards the OncoMouse as a being in which nature and culture come together, as her sister and sibling.

Recognizing that agencies are interconnected, that similar cultural patterns play out in beings of different species, and that there are different kinds of interdependencies between beings, is valuable. But when Haraway writes that the OncoMouse is her 'sibling' and 'sister', this is not simply an ontological statement: it is also a normative claim. She endorses animal experimentation by emphasizing the interconnectedness, but this conceals political human dominance. As Zipporah Weisberg writes, by comparing herself to the OncoMouse, Haraway sets up 'a false identity between herself – a relatively free and inviolable human subject – with a totally unfree and utterly violated subject-turned-object.' (2009, 49). Furthermore, there is no curiosity or invitation towards the actual OncoMice; they are not asked if they would consider that specific human kin too. I cannot speak for the OncoMouse, but it is difficult to see why they would consider Haraway as sister or sibling. Taking the OncoMice's perspective seriously demands a fundamental change in economic practices, political structures, and personal habits.

Words matter here too. Haraway uses a very different kind of vocabulary than Dolensek and his colleagues at the Max Planck Institute (2020). To those who are familiar with and sympathetic towards feminist, posthumanist, and new materialist theory, and acknowledge the need to rethink the figure of the human, many of her claims may sound convincing. But they are also deceiving, and a political lens shows they are inconsistent. Take 'Sharing Suffering', the title of the chapter of *When Species Meet* in which the OncoMice feature. Suffering is part of life for everyone, and we cannot keep others safe. Sharing suffering refers to empathy, recognition, and caring for others. I know what this means: when my dog or human companions suffer, I suffer too. But using this phrase in the context of the OncoMouse glosses over the large-scale human exploitation of mice. Perhaps individual workers can share suffering, but as societies humans should rather focus on not inflicting suffering. For Haraway, this framework of exploitation is legitimated by a 'greater good calculation' (2008, 87). But the calculation of this 'greater good' clearly depends on whom you ask – for the human greater good, sacrificing mice could make sense, but for the mice it does not.

Political and scientific structures are interconnected and mutually reinforce each other. In the production of knowledge, different agencies intermesh and affect outcomes. This insight challenges human supremacy and opens the way for acknowledging other agencies, ontologically and politically, which can be a basis for reforming relations. However, recognizing limits on individual human agency does not legitimate large scale institutional human violence. A political lens allows us to see the large-scale injustice that most people would not find acceptable in the case of humans. Finding it acceptable in the case of other animals is conceptually and morally problematic. Taking this seriously does not need to lead to abolishing all relations – as the new field of animal labor shows, for example, it can also mean reformulating them (Blattner et al. 2019).

What is a mouse? And who decides? After describing the lives of the songbirds she lived with in great detail, Len Howard (1953, 1956) concludes she cannot make general claims about bird species like great tits. Like humans, individual birds have very different personalities. While we can describe characteristics of species, and critically review how this relates to the roles they play in human societies and imaginations, this is an important point to keep in mind, especially for animals who are made invisible and replaceable in laboratories, and in different kinds of texts.

Learning to see mice differently

Learning to see other animals as their own beings takes time. Crary and Gruen (2022) describe how ideologies distort how humans view other animals. Portrayals of animals in documentaries and art can affirm or contest existing relations; seeing is not neutral but interconnected with animals' status in society (see also Pick 2011). When a human sees a mouse, what they see is inevitably shaped by received cultural opinions that follow from how they were socialized. As Crary and Gruen emphasize, humans are capable of moving beyond this immediate response, as they are in relation to other humans. This requires moving away from the self and focusing on the actual other; certain forms of art and critical theory can help to open the way.

Looking at the mice changed how I saw them. It took me several months to learn to tell them apart. The only one I recognized from the first day onwards was Vachtje, because her fur was a bit tousled. After a few months I began to perceive differences in size and body shape among the others. I began to notice that two of the mice were larger than average – Grote Muis and Kleinoor – while two were smaller – Kleine Muis and Kraaloog. Bullie was slightly bigger than all the others. Once I was able to describe everyone's posture, I could begin to recognize them as individuals. This took practice, and for a long time there were uncertainties, for example when someone was in the wheel and I could not tell if it was Grote Muis or Kleinoor, who look very similar except that Kleinoor's left ear is slightly smaller than the right one. But with time it became clearer, and I began to recognize their faces as well.

Barbara Smuts describes how learning to see other animals works in relation to a troop of baboons and a companion dog, Safi (2001). The processes she describes are very different – in the case of the baboons she needed to learn to 'speak baboon' in order to be accepted as a friendly stranger, but always kept her distance. In the case of Safi, the relationship includes close proximity and sharing a house, as well as developing daily habits. But in both cases, learning to see the animals correctly required an attentive awareness for Smuts, not just with her mind, but also with her body and spirit, that was fostered and learned in dialogue with the animals in question. Another example is the work of biologist Deborah Gordon (1992), who describes how watching ants for a long time allowed her to see them. Howard (1953, 1956) also emphasizes that learning to see other animals takes time. Observing animals in only one setting for a limited amount of time often leads to errors in the interpretation of their behavior, because one does not know the individual personalities of the animals involved, their habits, histories, and interpersonal relations. The mode of individualized observation that these researchers describe stands in stark contrast to the mechanistic laboratory setting in which individuals are only seen as representatives of their species, and allows researchers to ask very different kinds of questions.

Some of the acts of the mice were immediately clear to me, for example, when they were shy or curious. Other behaviors and activities took longer to understand, and the same applies to understanding their friendships, and the ways in which they create meaning more generally in terms of relations, play, spatial arrangements.

Learning to see mice is not just something that is interesting theoretically. It is also an important component of learning to care for them, on a day-to-day basis, and when they are ill. Because their lives are short, and there is not a lot of time to make mistakes. Through offering the mice different options concerning the houses, wheels, foods, spatial arrangements in their houses, and contact with me, I investigated their preferences. For example, their preference for holes in the ceilings of the boxes, or wooden wheels, may seem like a minor issue to a human, but it matters to them. What matters to them most is that I change the setup often, preferably daily. Trying out new things is fun for the mice, as they enjoy working on projects; but it is also important because they are curious, and investigating the new keeps them interested. Learning to see mice thus also involved a conversation, in which I learned from them through embodied questions and answers.

Seeing mice also matters in relation to knowing how to care for them when they are ill. Vachtje was the first mouse to fall ill, and the first to die because of her tumor. Breedsnuit and Mooitje also died from a tumor. Kleine Muis and Grote Muis both died rather suddenly, after being ill for only a day. Flankie and Kraaloog died of old age, though Flankie also had an ear infection and symptoms of paralysis in her hind legs. Health care for mice is not very advanced, which is ironic in light of the fact that we know so much about their bodies.

The bodies of lab mice are altered by humans for human purposes – sometimes genetically, sometimes they are made ill individually or collectively (Taylor 2017). When they live longer, many of these mice develop tumors. Because mice are seen as commodities and because they do not live long, their lives are not valued by human societies. There is usually little that can be done when they are ill – the vet's repertoire mainly consists of eye or ear ointment, painkillers, antibiotics, and euthanasia. I have been learning how best to assist them, and I have gotten better at judging situations. While it was easy to see that the mice were ill from the start, it took time to learn to understand how diseases progress, mostly with regard to the timing. Everything is faster with the mice than with cats, dogs, or humans. While my vet and the internet helped me judge the severity of situations, there is a certain knowledge that only comes with experience. This was not only true for me. The mice also learned about what it means when one of the others is ill, and have learned what death entails, as I discussed above.

An important aspect of learning to see others is allowing oneself to be transformed (Smuts 2001). Matthew Calarco (2018) argues that ethology should be seen as a transformative practice. He discusses the ethological work of Joe Hutto, who studied a pack of wild mule deer. Hutto was adopted by them and formed new social relations that made him see the reality of the deer differently, as well as making him more aware of the demise of the deer's lifeworld due to ecological collapse. Calarco draws attention to the social, environmental, and mental dimensions of ethology, which he describes as a pro-animal practice aiming to reform and re-envision relations between humans and other animals. Practicing ethology in this sense enriches one's world, but also makes one more sensitive to the suffering of other animals and the broader environmental catastrophe.

This is true for living with lab mice as well. Learning to understand the depth of their feelings and relations, caring for them, and being studied by them is joyful and a way of becoming more worldly. But it also makes reality more unbearable, because I now better understand the suffering of mice, inside and outside the laboratory. Experiencing the full cycle of life with these mice and witnessing their deaths, was also transformative. It made me view life itself differently, because it gave me a better understanding of its fleeting nature, and of how different the shape of one's life can be. Some mice suffered before they died, especially the ones who died when they were older. I looked after them and felt for them. I suffered, but did I share in their suffering? I am not sure. I do know that all of us are here only very briefly, and I learned that for the mice death matters exactly as much as for humans.

Political mouse-human dialogues

Working towards a better world for mice should be aimed at large-scale structural change. Current political, social, and economic practices and institutions can play a role in moving towards just interspecies societies. A first step towards more just relations with laboratory mice could for example be protection by labor rights, including the right to a pension. But change is also necessarily interconnected with small-scale experiments in which mice and other animals are consulted about their thoughts on the matter. Otherwise, humans still have the last word in determining what is good for others. There seems to be a large gap between the daily interactions that I have with the mice and larger political, economic, cultural, and social structures in which animals are exploited. However, there are different ways to bridge this gap. I will briefly explore two: viewing mouse-human conversations as components of larger systems of deliberation, and writing about mice.

Recent work in animal philosophy posits communicative interactions between humans and other animals as deliberative practices (Driessen 2014; Meijer 2019, Chapter 9; Donaldson 2020). Scholars of 'animal deliberation' (Driessen 2014) emphasize the embodied and habitual nature of human political conversations, as well as the need to analyze the role of power in constructing who has political voice (Meijer 2019, Chapter 9). While multispecies forms of deliberation already take place, they can be improved (Meijer 2019, Chapter 9, Haraway 2008). Multispecies deliberation requires taking seriously animal languages and multispecies language-games, as well as the material surroundings in which conversations take place. For example, Clemens Driessen (2014) points to the importance of material interventions for human-nonhuman animal deliberation, such as the cat flap and the milking machine – these give us something to speak about. Sue Donaldson (2020) directs the focus to the space in which humans and other animals deliberate. She argues that embodied interactions between animals of different species that take place in a shared commons can lead to more just relations between humans and other animals and new ways of co-existing. Through embodied interactions in which animal agency is foregrounded, new forms of government can come into being. An example of an organization working towards greater political equality in this way is VINE Sanctuary, an LGTBQ-led farmed animal sanctuary in Vermont. The human animals who reside there view the sanctuary as a multispecies community in which the other residents are not seen as beings with pre-determined interests, but as subjects who can co-create the conditions under which they live (Blattner et al. 2020; Gillespie 2018; jones 2014, 2023; Jones 2014; Meijer 2021a).

Within this framework, the interactions I have with the mice can be understood as ongoing conversations. Our dialogues are embodied, not dependent on human language, often revolve around objects, require curiosity from both sides, take time, will include misunderstandings and understandings, begin with a basic sense of connectedness as vulnerable beings, do not have a fixed outcome, and when there seems to be an outcome this does not mean one should stop being curious. I say something – perhaps by putting a willow ball in their house – and they tell me if they like it by using it or ignoring it or trying to get rid of it. Or I offer them small bits of vegan cheese and they will either eat it or not. Some of these conversations take place once or twice, others are ongoing. For example, I sometimes ask them if they already like my hand by putting my hand in their house, and they say no, again. Another example of an ongoing conversation is our music ritual. I know they like some songs for running so I play these, or invent new similar ones, to which they respond by running in the wheel or through their house. In these conversations embodied forms of language play a role, as well as music, habits and objects.

Recent proposals to consider the whole system of deliberation, instead of only considering the deliberative quality of separate spheres, offer a framework for translating the conversations that take place on the microlevel to larger structures (Mansbridge et al. 2012). Through connecting our deliberative practices to other political conversations, the dialogues between the mice and me could influence legislation and decision-making concerning mice, which could function as a first step in the transition towards a just multispecies democracy.

Our conversations can also affect cultural understandings of what it means to be a mouse. I give talks about the mice and write about their lives in newspapers and on my weblog, which may contribute to social and political conversations that take place on other levels of society, and more generally may invite other humans see them differently.

Meaning-making in the mouse house

The mice have different ways of creating meaning, in their relationships with one another, in relation to me, through using objects, and forming habits and rituals, like the party when their house is clean. Understanding their ways of creating meaning as a human requires paying attention to them, taking the time, and experimenting. It also requires thinking about language, in two senses. Better understanding the perspective of mice on our common world requires learning about their languages, and how these can and do co-shape our common world. But language also matters on the level of the words and stories I as a human use to think and write about them. It is important to be precise about actual interactions, but also to not use the type of clinical, generic language that is reserved for objects, which scientists now often use when writing about mice, because in that specific language-game there is too much that cannot be articulated, and it does not do justice to their ways of being.

On the day that the mice came to live with me, I gave them a large brown envelope. They were so happy with this envelope, moving in and out of it, and using it to hide and nap in. Their first running wheel was also a source of great joy. After spending the first six months of their lives in a small laboratory box, their world suddenly became much bigger and things began to happen. With more space and new objects, their relationships with each other could also change, and they continue to change now that they are old. In a society that respects mice their lives would probably have been better. But even under these circumstances, they created beauty and meaning, individually and as a community.

CONVERSATIONS WITH AMPHIBIAN AND HUMAN NEIGHBORS

'Hey little friend,' I say, and pick up a salamander from the road. The salamander looks at me. Their body is cold between my thumb and index finger.

A neighbor calls from across the street. 'Is it frogtime again?'

'Yes,' I call back, 'the rain woke them up.'

'It's so early,' she says. 'It must be because of the climate change. I will also keep an eye out for them.'

I thank her, and Doris and I bring the salamander to the pond.

Like every morning, Doris and I patrol the alleys behind our house. I lift the covers of the gullies to see if anyone fell in during the night. In the second one I spot a young male toad. I open the cover and slide my hand down the side into the water so that I can scoop him up without scaring him. I close my other hand around him. As I walk to the nearest garden where he can hide during the day, he wriggles, trying to escape. I put him down carefully, he crawls into the grass. Doris pulls on the leash, she wants to go home.

'Is it time for the toads again?' My neighbor J. sits in front of his house with his five-year-old daughter. I tell them it is, and they promise to walk around the pond in the evening. When I do my round, I see them from a distance: the girl has brought her horse on wheels and a friend, J. brought his partner, and their orange cat follows them. The children scream each time they find a frog, and J. uses his slipper to move them from the road to the grass near the water.

In the supermarket I meet a former neighbor. 'I read about your toad group in the local newspaper,' she says. 'What should I do when I see one on the road?'

The delivery man rings my doorbell. 'You are the frog lady, yes? There is one in front of my wheel.' He points to his white van.

'You have to watch them carefully,' I overhear one volunteer saying to another one. 'Look. That one is trying to go to the other pond. So you have to bring him there.' The other volunteer hesitates to pick up the frog, a rather large female. 'Don't be afraid,' the first one says. 'It feels scary at first, but you get used to it.' The new volunteer picks up the frog very carefully and brings her to the pond. A car stops for them.

Doris usually spots the amphibians before I see them. Volunteer I. brings her cat along when she does her rounds, he wears a little lamp on his collar so that the cars can see him.

'I felt a bit isolated because of the lockdown,' one of the volunteers writes to me at the end of the first season. It is 2021, and we had evening lockdowns because of Covid, in addition to the other pandemic restrictions. 'Walking really helped me, and it was nice to meet the other volunteers and do something useful.'

4. Assisting Amphibian Neighbors

Abstract: In this chapter I describe the embodied, material and spatial conversations that I have with frogs, toads, and salamanders, during their yearly migration in spring. I describe how my experience of place and time, and that of the other volunteers in my paddenwerkgroep (toad patrol), changed through walking the streets at night and engaging with the amphibians, and how this experience is co-determined by their agency. I argue that learning to see others is an embodied and moral process and explore what this can tell us about the embodied and spatial character of conversations more generally. I also investigate how the conversations with amphibians connect to dialogues with and between human neighbors, and how they can change the narrative of towns and cities.

Keywords: amphibian philosophy, amphibian agency, animal neighbors, animal geography, multispecies community

In February 2020, Doris, Olli and I moved out of the city to a small town just north of Amsterdam. The town is called Landsmeer (Land's Lake), it is located in between meadows and a nature reserve, and there is a lot of water everywhere – in ponds and ditches, and gardens after it rains, – most gardens are over a meter below sea level. On the first Sunday of March, the dogs and I went for the last walk of the day around 9 pm. It was already dark; it gets a lot darker here than in the city. When we passed by our neighbors' house, I saw a frog sitting on the sidewalk, in the light of the street lantern. I looked around and suddenly saw dozens of frogs and toads, on the sidewalks and in the gardens. The road I live on is quiet in the evening, especially on Sundays, but cars sometimes drive fast, and the bus also passes through it. All frogs and toads were sitting really still. I made sure not to step on them, but after a couple of meters I knew this was not enough. When amphibians come out of hibernation, they are still very slow, and need assistance in crossing the street. I saw someone who had just been killed in the middle of the street, and another frog who was sitting near them. The dogs and I had to walk onto the

---

Meijer, Eva. *Multispecies Dialogues: Doing Philosophy with Animals, Children, the Sea and Others.* Taylor & Francis Group, 2025
DOI: 10.5117/9789048564415_CH04

road – the dogs were not amused by all this, because they don't like too much activity in the evening – and I escorted the frog to the other side. I did not pick them up, I never picked up a frog before and did not want to hurt them.

The following day, I read up on the *paddentrek*, as it is called in Dutch, the toad migration. From late February to April, toads, frogs and salamanders come out of hibernation and go to the water where they were born to find a partner. The exact dates depend on the temperature and humidity: they wake up when it suddenly gets warmer, in combination with (a light) rain. On the days that followed, the dogs and I assisted everyone we saw on the road during our evening walk, I even picked some animals up because that was faster, but several toads and frogs were still killed by traffic. I knew that I had to set up a so-called *paddenwerkgroep*, a toad working group, for the next year.

In the summer that followed something else happened. When I walked through the alleyways behind my house, I suddenly saw two eyes looking at me from inside a road gully. They belonged to a small frog. I immediately understood that this frog was stuck. Fortunately, I could open the cover, and with my hand underneath the frog I could lift them up and out of the trap. I searched the rest of the gullies in the alleyway and on that first day managed to lift around twenty frogs and toads out of the gullies.

This practice became a daily routine, which takes around seven minutes. In the first summer I helped 182 individuals. I learned that frogs can survive for quite some time in the gullies, but that toads are vulnerable, so I need to go daily (salamanders rarely fall in; sometimes I find pregnant salamanders in the gullies, in early June). Because the frogs and toads are scared of my hands coming from above, I need to be careful. Toads are generally calm so I just move my hand underneath them, and they will sit still on it, but frogs can be very jumpy, so I need to either give them the impression they climb out themselves with the help of a moving hand-ladder, or close my hands around them. Sometimes they jump away and dive under or swim into the pipes. My technique is now quite good so this usually only happens when they are already hiding in the entrance of the pipe. Another thing to be mindful about is to make sure they do not jump out of my hand when they are on land, or jump back into the gully. I contacted the city council about this problem, but three years later we are still negotiating placing ladders.

Setting up the *paddenwerkgroep*

In January 2021 I began to look for volunteers to set up a group, with the help of local media. I had soon collected a group of fourteen humans, including the neighbors that live in the house in front of which I saw the first frog, and two women who were already assisting the frogs and toads in this town. I had almost no experience with frogs and toads, or with setting up this kind of group, so I contacted someone with a similar group in Amsterdam. They told me that they use Google Doc for making the schedule, but because the situation there is different – it is only one road, and we have to patrol around three ponds – they had no further advice. They were also unable to tell me when the migration would start. Finding out was mostly a matter of watching what the frogs and toads did and keeping track of the weather report.

On the eleventh of March, the days were getting warmer and I expected movement, so the first volunteers went out. On most evenings there were two of them, sometimes three, who walk a block of about nine streets, around the three large ponds where most amphibians travel to, from an hour before sunset to an hour after sunset. It was moving to see the volunteers out in the dark, with their flashlights and the concentrated looks on their faces. I went out most nights as well, together with Doris. I also bought traffic signs, and drew frog faces on the road with chalk, accompanied by warnings for drivers. Some children joined the group as well, they were very good at spotting the amphibians, and one boy devised a ladder that he used to help the frogs climb out of the gullies near the children's playground. Interestingly, the frogs immediately knew what he meant. In that first year, together we managed to help frogs, toads and salamanders cross the street safely 586 times.

The group is now three years old. The number of volunteers grew, and the group became more of a community. The project affected others too: some neighbors come out to help on busy evenings, often with their children or grandchildren. In 2022 we assisted amphibians 1103 times, in 2023 we assisted them 1714 times, and in 2024 over 2000 times. The project changed something for the volunteers, including myself, and for the town. We became more attuned to the landscape and the animals, and the status of the amphibians changed because humans now helped them. The perception of the volunteers changed through walking, and seeing and touching the amphibians. The project involved conversations with amphibians and led to further conversations between humans, about the animals, the weather and the environment, which in turn led to a change in the human community.

Seeing and touching toads, frogs and salamanders

The conversations I have with frogs, toads and salamanders are very different from those with Olli or the mice. They are as brief as possible (because the amphibians do not like to be touched) and do not involve getting to know each other. While I often greet the animals with my voice, the main interaction we have is physical: I touch the animals, and do not ask for their consent beforehand. I do take care to approach them in a gentle way, and pick them up and put them down similarly. The process roughly goes as follows.

If I enter a street, I scan the road with my flash light. If there is someone in the middle of the road, they are the priority, so I first move them. I then look at the sidewalks and the gardens, and assist animals waiting to cross the street there. I also look out for cars, bicycles and humans on foot, and warn them if they approach someone – many humans do not look where they walk. When assisting an amphibian, the first thing I look at is the direction in which they are moving or facing. I have learned that it only makes sense to bring them where they want to go – otherwise they will cross the road again to move to their desired spot. I also judge the speed with which they are moving, in relation to possible cars.

When I see a frog, toad or salamander I always pick them up very gently. Toads, frogs and salamanders are not cold, slimy, or passive. Their bodies are very soft and all individuals have their own response to being picked up, so each encounter reveals something about the person in question. Similar to how this works with birds, the trick is to be very gentle when you pick them up, while still making sure they cannot jump out of your hands because they would hurt themselves. I often pick up toads with my thumb and index finger beneath their armpits gently, after which I put them in my other hand and close my fingers around them. Frogs are usually more feisty, so I use both hands to scoop them up and make sure to keep my fingers closed firmly when I walk to the water side because they will try to wriggle their way out. Salamanders are easy because they cannot jump or run: I simply pick them up between my thumb and index finger, and bring them to safety. They are the most vulnerable group of the three because they are small and very slow. The difficulty with them is not to catch them but to see them. Through looking at frogs, toads and salamanders, I learned to notice them; through touching them I learned how to best touch them. Both processes contributed to becoming more attuned to them, and to the larger environment.

Frogs and toads have a specific way of sitting, like little Buddhas. Toads crawl in a certain way, frogs jump. The salamanders are a long shape on the road or sidewalk. These basic images stand out from the background if you take the time to look for them, and through looking for them several days in a row, you begin to see them more easily. Your eyes learn to do the work without thinking, like your fingers can learn to play music without thinking about the notes. I discussed this process – learning to see the animals in the dark – with some of the volunteers, who had similar experiences. In addition to learning to see the animals themselves, one also learns where they like to sit, for example in the shadows of sidewalks, or the alleyways that lead away from the gardens, which makes it easier to find them. Because it is dark when we walk, seeing requires attention (and a good flashlight). There is a game-like quality to the walk – the walkers want to find as many animals as possible and email me their numbers with pride. In the beginning of the season there is also always a sort of frog fever, connected to wanting to find as many as possible – are they out tonight? Will I find them? While this helps the volunteers to stay interested, this focus on the grandiose also has a downside, which is that they are disappointed when they do not find many animals, so I have often tell them that every life counts. But there is also a more meditative aspect to the practice. Looking for amphibians involves quiet concentration. Sometimes I walk for two hours in rainy and cold weather and only find two animals, who might have very well crossed the street safely without me. This still contributes to learning to become more worldly.

How I touch them is formed by my experience. In the beginning I was very careful and sometimes let them escape, I also once let a toad jump out of my hands. Fortunately he survived. Now I pick them up quickly and bring them to the water in seconds. My grip is sometimes loose and sometimes tighter, depending on the animal in question. In conversations with me, the volunteers described being hesitant or even scared when they first picked up a frog or toad. I still feel this sometimes with really big wild frogs – there is something strange about them, something other. While I was never scared of them, picking up amphibians did feel unfamiliar, and now it is a familiar practice: my body knows what it has to do and acts quickly. Ike Kamphof (2024) pointed out to me that the experience of touch of humans and amphibians during the migration is different: humans like to touch amphibians, while amphibians are scared of humans and would prefer not to be touched. I keep this in mind when touching them and teach the volunteers this too. Still, engaging with them changed my embodied experience of our shared environment.

Becoming more attuned to the amphibians led to other sensations too, such as a sense of pride when I hear them croak in spring or see the spawn in the ponds, and a sense of satisfaction when I hear the occasional 'plop' of them jumping into the water. I worry about them when I am away in the evening during frog season, and I miss them in the winters when they go into hibernation and I do not have to go on my daily round through the alleyways anymore.

Through caring for these animals, and working with the human volunteers, some of whom want to share their personal problems with me too, I became more rooted in the town. In part, this is a social phenomenon: I have not made human friends through the group, but do speak to many of the neighbors regularly now, and am always happy to meet my amphibian neighbors. But it also changed how I am situated in space and time, in this town: I read the seasons, the weather and the time of day differently, and experience the streets as multispecies spaces (in which traffic and humans are dangerous for amphibians). One significant change in me was learning to read the weather. Barbara Smuts (2001) describes how walking with a troop of baboons led her to experience her surroundings differently, and taught her to predict changes in the weather. The baboons always knew the exact moment to move when they were eating in a field and saw a rainstorm approaching. For months, Smuts wanted to run long before they did, but through watching the baboons and moving with them, she learned to read the weather too. Not with her mind, but immediately, physically. She describes this as a 'small but significant triumph' (2001, 299), because she went from thinking analytically about experience, to experiencing directly, and she describes the baboons as the experts who guided the way for her. Through frog walking I experience something similar. In the first year of the toad group, I judged if nights would be busy or not on the basis of information from the internet and by making lots of extra walks. Now I usually know. It's a combination of temperature, humidity or rain, and wind – I also found that they are active on windy days, contrary to what the internet says. The amphibian migration marks the movement from winter to spring. The salamanders, who mostly live around the small children's playground near the third pond, are always the first to move. In the year, they are sometimes two weeks earlier than the frogs. Toads follow, frogs are later (frogs are also later in the evening; when the toads walk around 7 pm, the frogs arrive around 8 pm). These animals help me to experience the changes in time more consciously, which changes how I live in time.

Learning to see others differently

Walking, looking for amphibians, picking them up and bringing them to the water, is a practice in which humans learn to become more attuned to these animals and the environment. This is an embodied experience, one that conversations with humans can help to make sense of. Learning to pay attention to the frogs, toads and salamanders also has a moral side: the volunteers already felt that the lives of these animals have value, because they otherwise would not have bothered to join the group, but through helping them this became much more visible and important to them. After a few weeks, new volunteers usually email to tell me that they find it difficult to see the dead frogs and toads on the road.

Philosopher Elisa Aaltola (2019) draws on the work of Iris Murdoch and Simone Weil to argue that we need to cultivate attention in order to develop a proper moral attitude towards other animals. Both Murdoch and Weil see attention, understood as an expression of love, as a way of getting closer to the truth. Aaltola rejects the idea of truth as universal and absolute in the Platonic sense, but argues that attentive love, understood as a focus away from the self towards a particular other, can help us see more clearly. Furthermore, learning to attend to other animals can help us to move beyond the anthropocentric bias that leads to much of the violence present in the world today. This works as follows.

According to Murdoch (1971), western philosophy has mostly focused on the will in thinking about morality, not on vision or attention (Aaltola, 2019). This has led to a loss of world: by focusing on one's own position or inner state, humans become detached from their surroundings. This focus on the self also lets the realities of others fade into the background. Similar to Buddhist ideas about the self, Murdoch sees the ego as an illusion that veils the world. Murdoch writes about how this affects individuals; Aaltola argues this happens on the level of our species too. Because of a strong focus on their own species, humans see other animals only or mostly in relation to their use for humans, and lose sight of their lived realities.

As a remedy, Murdoch proposes a move away from the self. Crucial in unselfing is paying attention to specific others. Through attending to what is outside of you, you come to see the world as it is (Murdoch 1971, 89). Attention is not grand or spectacular; rather, it is connected to focusing on the daily realities around us. By looking at a bird, tree, or rock, one can practice unselfing. '[The particular] teaches us love, we understand it, we see it' (Murdoch 2003, 497). Importantly, learning to attend to what surrounds you is a practice that takes place over the course of a lifetime, it is not a skill one learns once and for all. Such attention is love, according to Murdoch: it is the work of love to attend to others, and attending can lead us to connect to the realities of others, beyond our prejudices about them (Aaltola 2019).

The connections drawn in this way between attention, learning to see, and morality can shed light on some of the processes I described above. Through practicing looking at frogs, toads, and salamanders, the volunteers and I began to see our environment differently. Through actual embodied interactions with them – touching them, trying to understand where they want to go, making sure they are safe – we encounter their realities, and understand certain aspects of it. This is tied to understanding that humans are not the only ones who live in this town. We share our gardens, alleyways and ponds with many others. I suppose many humans would agree with this idea, at least in an abstract sense. But through attending to specific others, we become part of the same shared world.

Looking for frogs, toads, and salamanders is a meditative practice, in which attention to others helps to set aside the ego. During the walks the attention is not on oneself but on the ground, and one really needs to pay attention, because it is dark and otherwise you miss the amphibians. There is also no way of rescuing everyone, so during the walks you will encounter dead bodies, and perhaps wounded people, which puts one's effort (and being) in perspective.

In thinking about animals and attention, Aaltola rightly writes that we should watch out for egoistic forms of love, that affirm the apparent good character of the 'I' in question. For example, humans might love their dog companions and think they are good to animals while still eating pigs or fishes.[1] This danger is also clearly present in the case of liminal animals. Humans might watch and feed songbirds in their garden and still eat chickens, and see these animals as belonging to different categories.

I do not ask the volunteers about their eating habits or general views on animal rights, but I am quite sure not all are vegan or invested in the project of animal rights (two of them volunteer for the animal ambulance though). While I know that their attitudes towards the amphibians change through walking, I do not know if this translates to other animals. Anthropocentrism, however, not only shows in practices of domination, such as eating animals. It also informs how we consider spaces. As I will discuss in the next section, the project contributes to sharing space differently with other animals, and learning to attend to nonhuman neighbors. Again, many of the volunteers were already sensitive to the idea that we share this town with other animals. But through investing time and attention, the lived realities of some of these other animals became clearer, as well as the obstacles they face to lead a good life.

Learning from amphibians

The agency of the amphibians is central to learning to see them, and to the practice of assisting them. When new volunteers sign up, I send them

1 Ethologist Jonathan Balcombe proposes to use the word 'fishes' rather than 'fish' to acknowledge that these beings are subjects with their own personalities and relationships, like other animals.

a document with information, which includes pictures of the different species we assist, instructions about handling them and preparations (like washing your hands), a map with the streets we patrol, and information about times. On the first walk, new volunteers are coupled with an experienced volunteer (which is often me). But the actual learning takes place through doing, and in this process the agency of the amphibians matters most: we need to follow them, both in where they want to go and how they want to be handled. Human instructions matter, but are not sufficient to learn to walk well. Furthermore, through following the amphibians, we encounter their lived realities, which changes our perspective of them, and the place in which we walk.

Philosopher Jeff Stickney (2020) highlights the importance of the environment in which we think for the quality of our thinking.[2] Environmental education that takes place outside and connects specific natural beings and objects to philosophical reflection can change the grammar of a place. Stickney recognizes that there are already different grammars at work in different areas – in the context of Toronto there are, for example, indigenous grammars of place that are different from settler narratives when it comes to trees or ecosystems – but he focuses on human grammars, and on human educators in the movement towards seeing what surrounds us differently. As the toad group shows, the amphibians play an important role in portraying their position and the town, our shared reality, differently. For the humans involved, this changes the grammar from human to multispecies.

Understanding how this works requires attending to language beyond human words, and embodied modes of learning and teaching. Learning to see the animals involves training, learning to pick them up involves touching.

2 Stickney develops an idea of place-based environmental education through philosophical walks. Taking students outdoors to discover and discuss nature has been a longstanding tradition in the history of philosophy, popular in ancient Greece, and most likely pedagogy before that. Stickney emphasizes the relevance for the outdoor classroom, which has been seen as relevant for different reasons by different philosophers, in thinking about nature, animals, and the climate crisis. He describes a walk, in which there are different moments for reflection. Specific points in the environment – a tree, or object – are connected to larger reflections. After the walk students discuss what they saw, for example through mock deliberative fora. A sense of wonder is central to the questions Stickney discusses – about tree communication or the stars – and a connection to the specific space. Teaching students about connections in the landscape and between humans and the landscape asks the teachers to be immersed in the place, and to take the walks often: this way they can develop a 'narrative web' (2020, 1074) in which the place functions similar to an ecosystem or community. These kinds of walks can be used to teach students about embodied thinking and becoming more worldly, which can address some of the problems I raised in the introduction about modern academic philosophy. They can be part of multispecies education as well (see Chapter 8).

The amphibians teach us by moving in certain ways, responding to touch, and being determined to reach their destination. All of this contributes to seeing them differently. Stickney does not discuss the embodied dimensions of his educational walks in detail – he mentions biology students conducting experiments, but focuses mainly on verbal transmission of knowledge. But being outside does not just change our view and thoughts: it affects our whole body, and with our bodies we affect the land – our footsteps, breath, the sounds that we make, affect our surroundings. Walking itself plays a role too in how and what we learn. Research shows that walking changes what we see visually and enhances the peripheral visual field (Cao and Händel 2019); it also affects how we think (De Certeau 2014).

Learning to see the amphibians differently changes the humans involved in the group, but the project affects larger society too: the humans with the flashlights have become a familiar sight for those living in the streets where we walk, and through interviews with the media others have become aware of the fact that amphibians migrate and might need assistance.

Living better with amphibian neighbors

The main message that the group conveys is that the amphibians' lives matter and that this is their town too. Philosopher Ralph Acampora (2004) turns to the figure of the neighbor in order to clarify the moral obligations towards the animals with whom humans share their living areas. Our compassion is, and should not be, reserved for humans in our proximity, Acampora writes, but also for other inhabitants of the earth. Sometimes these relations are close, as with companion animals, but most are not. Still, our lives can be connected in important ways, for example when we share a garden with frogs and toads. Acampora writes that the figure of the neighbor captures this proximity or even intimacy, and sees it therefore as more fitting for thinking about multispecies relations in urban areas than, for example, political concepts as 'citizen' or 'denizen'. While I disagree with Acampora about this latter statement – which conceptual lens we should use depends on the questions at stake – I do think the figure of the neighbor offers a good way of thinking about reforming relations with the amphibians. Viewing them as neighbors shows the importance of sharing the land justly with them, because it is theirs too, and respecting their way of life. The precise obligations of human neighbors towards their nonhuman neighbors differ, based on our responsibility, history, and (in)dependance (Acampora 2004, 230).

The frogs, toads and salamanders we assist do not depend on humans for food or care, they only want to cross the street safely. Our duties towards them are relatively light, compared to companion species, or animals who are more strongly affected by human acts. However, taking them seriously as neighbors would require a ban of buses, cars and bicycles during dusk and an hour afterwards in the migration period, and perhaps assistance on the two or three busiest nights of the year to make sure that humans do not step on them. It would also ask for protecting their habitats, because they now depend on the goodwill of individual humans for keeping their hibernation areas intact (like gardens: building sheds in winter can destroy many lives), and placing ladders in gullies. Since they are not yet recognized as neighbors by most humans, helping them cross the street is the best option, and one that can contribute to seeing them differently, through changing the narrative of the town and spatial interventions.

Changing the story of the town through spatial practices and conversations
For many humans who grew up in this neighborhood, the amphibians belong to the town. While they do not care for them, they do know about the migration, and for some that I spoke to, it signals spring. One of the volunteers, who is in her seventies, has told me that the amphibians have been following the same routes since the 1950s.[3] Since then, new houses have been built, there is much more traffic – cars and buses are the biggest problems, but bicycles also kill amphibians and sometimes humans step on them – and there is an ongoing trend to tile gardens. The frogs, toads, and salamanders have some room to respond to these changes, they can move to other gardens, and even meet companions in other ponds, but they cannot adapt to the traffic, the gullies, and changes that happen when they hibernate.

For neighbors who moved to this town later in life, the amphibians are often invisible – my own neighbors, who joined the group and have been living here for over twenty years, admitted to never having noticed them before I pointed them out to them. Furthermore, even those who did grow up with them often do not notice them on the streets when they drive or walk, and their dead bodies on the road do not cause them distress. The toad group counters this narrative of disinterest through showing we care by walking. The amphibians move regardless of humans noticing them or not,

3 In the first written account about their habitats in the Netherlands, Gronovius (1756) writes that toads, brown frogs, and water salamanders often live in gardens (Van Diepenbeek et al., 2009). While there are no written accounts of frogs, toads, and salamanders in our town from that time in the archives, it is plausible to assume that they have been around for centuries.

but the working group amplifies their movement by adding their movement to it. This makes the annual migration more visible, and shows that there are humans who care, which affects the moral views of at least some human neighbors about the amphibians, and changes the story of the town.

In order to understand how the story of a place can change, Simonsen (2008) develops two analytical figures: the embodied city and the narrative city. Simonsen proposes to view the city as an embodied meeting space. Urban life is formed by moving bodies, which carry power structures, norms and memories with them that are developed in specific places, and which change these spaces creatively by their acts (through appropriating spaces or connecting them). The city also holds together a narrative: collectives and individuals live out different stories in relation to the specific city surroundings, which they also affect. The narratives that come into being are 'spatio-temporal operations connecting future, past and present' (2008, 146). These form memories and expectations, shape conversations, and are shaped by them.

In exploring the multicultural narrative of Copenhagen, Simonsen draws attention to how human positionality is shaped by colonial practices. The legacy of colonialism leads to difficulty with encountering difference in western parts of the world, and shapes a narrative in which some belong and others not. Some spaces in Danish cities are 'othered', for example when they are called ghettos, and this draws boundaries between groups of humans. A second perspective shaping the narrative is cosmopolitanism, which Simonsen describes as a narrative that emphasizes all humans are members of the same world community. Both positions play a role in public discourse, and inform everyday constructions of the multicultural city.

The idea of the embodied city and the narrative city, as well as the influence of ideological structures on social narratives, can help us see how the toad group changes the narrative of the town. The group intervenes in the usual business of the embodied city, in which humans generally carelessly run over migrating animals with their cars, bicycles, or feet, by acting differently, walking and picking up animals, and showing that there are more bodies than only the human ones. This presents a different human story, but also alters the narrative of the amphibians, who suddenly have a much larger chance of surviving, and who have a different kind of contact with humans (which they do not appreciate, so the narrative is different for the amphibians than for the humans).

While our positionality is shaped by anthropocentrism, which others nonhuman animals, there is also a counternarrative in which they belong to the town, and this latter narrative is strengthened by the acts of the group. The group is still relatively new, but responses to the project have changed. In the first year, we constantly explained what we were doing, and now neighbors know. They still come up to us for information or help. Because the project gained visibility through interviews with newspapers, I was also approached by two humans in nearby villages who want to set up similar groups in their area, and have assisted them in setting up their groups. However, the most important signal of change is that many neighbors join us for one or two nights, which makes the streets come alive. Last year, a large group of neighbors joined us on the busiest nights of the year, including many children. Their voices, the festive atmosphere, and the flashlights made it look like a new spring ritual.

Finally, learning to be better neighbors to amphibians is, again, also a spatial matter. In a discussion about the value of urban wildlife, Diane Michelfelder (2003) draws attention to the role of design and architecture in improving multispecies relations. Proposing to view urban wildlife not just as aesthetic object-like beings but as a component of larger urban community – indeed, as neighbors – she offers the idea of designing social spaces for multispecies interactions. The gardens could be starting points, but the streets too – making them car-free for a period of time each year (or always) would allow for new encounters and interactions to take place. Furthermore, spaces could be designed differently to make the lives of other animals safer. In some Dutch spaces design and architecture already play a role in working towards new urban multispecies forms of co-habitation in relation to amphibians. For example, in The Hague, the local branch of the Party for the Animals has established that all road gullies in the city now have ladders or nets that make sure amphibians do not fall into them and get trapped. This type of intervention is fairly easy to establish and largely goes unnoticed by humans, but it matters a great deal to the animals – no large-scale studies exist but it is estimated that over a million amphibians die in gullies yearly, many of them toads. Another example is the ecoduct: tunnels under or bridges over highways that make it possible for amphibians to cross roads safely. Making these designs visible in a town also helps humans see other animals, and contributes to changing the narrative too.

Spatial conversations and worldliness[4]

The conversations between humans, and between humans and amphibians I described above, share some of the characteristics of the dialogues I

4 The passage about fūdo appeared in a slightly different form in the essay 'Missing Winter' (2024), in the journal *oxford public philosophy*.

discussed earlier. They are embodied and extend beyond human language, and they involve change on the side of the humans. They further show the importance of space as well as time. The conversations are not only shaped by the bodies and minds of those involved, by also by the weather, and the material surroundings such as roads and gardens. They take place in a specific part of the year, and on a specific part of the day.

Through walking, I became more attuned to the animals, but it also affected my awareness of being in a certain place and time, like when I learned to read the weather. The Japanese philosopher Watsuji Tetsuro (1961) captures the subjective embeddedness of humans in their environment with the term 'fūdo'. Fūdo literally means 'Wind and Earth' in Japanese, but Watsuji uses it to refer to 'the climate, the weather conditions, the nature of the soil, and the geological, topographic, and scenic features of a given land' (Johnson 2018, 1134). Watsuji develops this idea of fūdo in response to Heidegger who, according to him, focuses too strongly on time, at the cost of our spatial existence. Time and history are, in Watsuji's view, always embedded in a specific fūdo. This is expressed in the experience of humans, who are not loose subjects who accidentally roam around in a given environment, but who are shaped by it. Watsuji connects this emphasis on locality to being with others. If we emphasize our being in time, we understand human existence on the level of individual consciousness (Watsuji 1961, 9). But according to Watsuji, 'being-in-relation-to-others [...] is the essential place of standing out (ek-sistere)' (Watsuji 1961, as cited in Johnson 2018, 1138). Ek-sistere, for Heidegger, refers to existing in time, which is dynamic and moves towards the future. But understanding our being as Being in space too, including the spatial dimensions of time that are expressed in history and culture, reveals that we are fundamentally social beings. This is ontological for Watsuji, it is part of the kind of beings that we are. As embodied beings, human are formed by and form fūdo, which is then actualized in *fūdosei*.

Ralph Acampora (2006) points out that we share fūdo, which he translates as 'climaticity', with the other animals. In general, because fūdo is a characteristic of all animals, similar to vulnerability and mortality, but also in given environments, where we may share a fūdosei. Helping the frogs, toads, and salamanders cross the road connected me to the weather and the beginning of spring with my body – the amphibians come out of hibernation when it is warm and rainy, and being attentive to the right conditions made me find a home in this town in a different way than before. Like the amphibians, my days and experiences are shaped by the conditions of this watery town, and becoming attuned to them made me aware of the fūdosei we share.

At the same time, we do not all experience this fūdosei similarly. For example, the amphibians have a different physical understanding of the climate, air, and water. The amphibians can breathe through their skin, they metamorphize, they lay many eggs, salamanders change color in the water – all skills that humans do not have. Due to their size, they experience traffic very differently from humans. They experience our assistance differently: humans in the group like picking up amphibians, but they are generally afraid of humans. Of course, species is only one factor amongst many, and species membership should not be overexaggerated when it comes to understanding others or creating meaning together. But differences matter too, and can provide us with a richer understanding of the possibilities of relating to the environment in which we are embedded.

Attending to the amphibians and becoming aware of our common fūdosei can guide the movement from world to worlds that I mentioned in the introduction: for me, it was a way of becoming more worldly. Of becoming more of the world, through contributing to a specific life-world in this world. The habitats of humans and amphibians are under threat because of the climate crisis, and related crises. As individuals we cannot turn this around singlehandedly, because large companies are responsible for most emissions. But we can organize politically and strive for cultural change, and we can attend to the worlds of others and co-constitute new life worlds with them. Tending to life-worlds, and co-forming new worlds, is not a solitary project: other animals and plants are constantly in the process of creating meaning, together and in relation to humans. Kim Tallbear describes this coming together of experiences with the words of Vine Deloria Jr. as a 'social reality, a fabric of life in which everything had the possibility of intimate knowing relationships because, ultimately everything was related' (TallBear 2011, 3). Different ways and modes of knowing do not obstruct relations, but form and inform them.

Taking seriously the world of the amphibians has made me see my own position in life differently. 'After a while, being with them felt more like "the real world" than life back home,' Smuts writes about living with baboons (2001, 299). She connects this to learning to be more animal, and letting go of human layers of thinking. Attending to the baboons helps her relate to her environment and life differently. Assisting the toads to me often also feels more real than many of the human projects I take part in. Not exactly in the sense that Smuts describes – while I do become attuned to them, I do not share their life, and our encounters are fleeting. The feeling of realness is connected to seeing their reality, and the sense of urgency connected to helping them. Assisting the amphibians often feels like a race against the clock, because of cars and buses, and the fact that you cannot be everywhere at once. Patrolling also means seeing many dead bodies, and, in my case, killing those who are mortally wounded but not dead yet – health care for amphibians is non-existent. Every year I kill around ten animals, and I carry the memories of these acts in my body. This too changes how I think about life, and my position in it. While I share an environment with these amphibians, I also have duties towards them. As a neighbor, and as a fellow being.

Conversations with amphibians

Dialogues with dogs and mice share a family resemblance with dialogues between humans. They include many of the same language-games, even though their shape might be different. Think of greeting, for example, or asking a question. These language-games do not rely on human language, but nor do human language-games. Speaking with these animals requires paying more attention to the embodied and material aspects of the communication, but in speaking with humans this also improves interaction. Of course, there are differences in what we can share with others of different species, depending on physical characteristics, social relations, and personal histories. At the same time, in the human case there is also a large variety in physical and mental capacities that affects which language-games we can share with others, and what we can share is shaped by social relations and personal histories too. Furthermore, there are language-games we can share with nonhuman animals and not with humans; think, for example, of writing with a cat on your lap or a dog touching your side on the couch.

With the amphibians, the situation is different, because I did not develop close relations with them and because their way of being is different from mine in another way than the dog way of being is. They hibernate, live on land and in the water, have different kinds of relations with others. However, encounters with amphibian neighbors share characteristics with those with human neighbors, and in the context of the town different humans and amphibians all have their own projects and try to make the best of their lives. The volunteers of our group are in conversation with the amphibians individually, in direct encounters, and collectively, through a more symbolic process of question and response. The amphibians do not ask for help by means of a representative of their community or in human words, humans respond to their vulnerability out of their own accord. Viewing these collective spatial interactions as conversations stretches the meaning of the concept 'conversation'. However, in the case of humans too we find similar material large-scale conversations between communities and even nations, that may involve direct encounters but also interventions such as building walls or fences, or impressing others by showing off. Still, extending the concept of 'conversation' in this way does require exploring the boundaries of conversations in more detail. In the next chapter I continue this by examining conversations with the North Sea, and its inhabitants.

CONVERSATIONS WITH THE SEA

In January 1647, a large storm near Domburg in Zeeland, The Netherlands, brought ashore stones with half-weathered images and thanksgiving prayers to a goddess named Nehalennia. These were likely remnants of a temple, including petrified and preserved remains of trees in salt. At the time, no one knew who Nehalennia was, and the finding caused great excitement among scholars, who thought of it as the sea returning something unknown to the people (Wagenvoort 1971).

In 1961, Dutch artist Wim T. Schippers emptied a bottle of lemonade in the North Sea, in a performance called *Flesje limonade gazeuse in zee bij Petten*.

The greatest natural disaster of the twentieth century in the Netherlands was the *Watersnoodramp*. The North Sea flooded Zeeland in the south of the country in 1953, killing 1836 humans as well as an unknown number of other land animals. This led to the Delta Plan, a series of construction projects in the southwest of the Netherlands, including dams, dykes, and storm surge barriers.

When I studied art in The Hague in the early 2000s, Pika and I often took long walks on the beach. We took the bus to Kijkduin, then walked through the dunes and followed the beach back to our starting point. We preferred cold and rainy weather, when it was just us and the sea. Pika liked to swim in all seasons, she was very strong. Looking back, the memories of these walks are the best memories I have of these years. There is a sense of belonging in them, because we were together, and felt at home near the sea.

5. Conversations with and about the North Sea

Abstract: This chapter explores the limits of the concepts 'dialogue' and 'conversation' through an analysis of conversations humans have about and with the North Sea. I compare conversations with animals to those we have with plants (notably seaweed), epistemologically and normatively, using a vegan ecofeminist lens to bring into conversation scholarship in critical animal and critical plant studies. I argue that plants and animals should have a right to speak about our common world, in the context of the sea. I also explore how we can engage differently with the sea itself, by looking at existing examples of humans who speak with the sea, and analyzing these using western and indigenous perspectives. Viewing interaction with the sea as a set of conversations can be of normative importance and change the attitude of humans. It also directs our view to how human language works and asks us to be attentive to the varied meanings that concepts can have in different situations. Learning to speak better with the sea also requires acknowledging that human agency is always part of a larger network of relations.

Keywords: sea philosophy, agency, seaweed, water philosophy

I grew up in Hoorn, a town close to the Markermeer in the north-west of the Netherlands. Like nearly all Dutch children, I learned to swim at a young age.[1] In winter my sister and I skated on the frozen canal at the end of our street. Water is never far away in the Netherlands, with all its canals, rivers, *slootjes* (ditches – ditch jumping is a popular pastime for youngsters), lakes, and the sea. Humans who grow up here learn to enjoy its presence,

1 School swimming lessons are supported by the government, and nearly all schools offer them, but since swimming is such a popular activity in summer, especially for children, many children learn the basics from their parents when they are toddlers, or at the very least become acquainted with water at that age. I have no memories in which I cannot swim.

Meijer, Eva. *Multispecies Dialogues: Doing Philosophy with Animals, Children, the Sea and Others.* Taylor & Francis Group, 2025
DOI: 10.5117/9789048564415_CH05

but also to fear it. When I was young, reckoning with the force of water was a simple, straightforward fact of life: do not go too deep into the sea, the currents might be too strong there.[2] Or: do not take your rubber boat out too far onto the lake, you will not be able to row back. But things have changed. The canal does not freeze over anymore in winter; my sister's children grow up without ice and snow,[3] and the water is becoming a threat again. The stability that the Dutch dikes provided for living beneath sea level safely for so long is challenged by expected rises in that sea level, and especially storms. Traditionally, the Dutch prided themselves on conquering the water and the ideal of cultivating and taming nature and the landscape still governs many political choices (Helmreich 2023). But the climate crisis and its floods, storms, and dry summers are slowly but steadily pushing the Dutch to think and act differently: not against, but with the rivers and seas that threaten towns and cities.

How to think with other earth beings, instead of just about them, is currently being explored in many different fields of study. The views that humans in western knowledge paradigms hold about animals, plants and the natural world, are changing. In the introduction I wrote that this follows in part from new research about their capacities, which raises normative questions concerning human duties and forming new relations. It also follows from the fact that the climate crisis and related ecological disasters press humans to rethink their position in a larger more-than-human world. Examples of this changing of views are the growing attention for rights for nature (e,g, Tanasescu 2022; Biemann and Tavares 2014) and multispecies justice (e.g. Celermajer et al. 2020; Donaldson and Kymlicka 2011; Lawrence 2022; Tschakert et al. 2021). Of course, viewing humans as part of nature – and challenging definitions of nature that exclude humans – is not new (Winter 2023). Indigenous thinkers (Kimmerer 2013; Simpson 2015) point out that the opposition drawn in western frameworks of knowledge between 'nature' and 'human' is not necessary, and not universal: many cosmologies and epistemologies see humans as part of a larger living whole. As animals we are part of a larger living web, with plants, the land, air, and water. With regard to the sea, there are many mythologies and cosmologies in which humans and sea are kin (McNiven 2016).

In this chapter I explore how the model of the dialogue can help us think and act differently with water, focusing specifically on the North

2 The depth of the sea is measured on the body: go no further than your waist.

3 I miss the snow. In the essay 'Missing Winter' I describe the melancholy that follows from this loss (Meijer 2024b).

Sea, the largest natural entity in the Netherlands. I consider two types of conversations. In the first half of the chapter, I focus on conversations with nonhuman beings about the sea, and in the second half on conversations with the sea itself. In contrast to the last three chapters, I do not draw on my own experience with the North Sea, but rather investigate the different types of dialogues that take place in and around the sea on the level of society. I also do not discuss one dialogue in detail, but I instead map different sorts of conversations, to get a better grasp on what 'dialogue' means in relation to more-than-human beings and natural entities, and to investigate how we can move forward in theory as well as in practice. This selection of dialogues is not meant to represent all types of conversations there are, with and about the sea. However, the conversations I discuss do give an impression of the most important questions at stake, and taken together they sketch a landscape, or rather, seascape, of dialogues.

An underlying aim of this chapter is to investigate how we can conduct different conversations at the same time, and how conversations with different groups – like fishes and seaweed, or waves and wind – relate epistemologically, morally, and politically. Often, scholars focus on either animals (e.g. Donaldson and Kymlicka 2011) or plants (e.g. Hall 2011; Lawrence 2022, Marder 2013, Sandilands 2021), but better understanding human agency and human duties towards others asks for mapping relations with different actors simultaneously, and looking for similarities and differences (Gaard 2016). Mapping relations with different actors matters because human acts generally affect many beings at the same time, so judging the violence of existing situations and articulating alternatives to the status quo asks for taking different beings into account. But it is also necessary get a full, or at least a better, view of our topic of conversation, in this case the sea.

Speaking with sea animals and plants about the sea

When humans think about how to relate to the sea, they often turn to human stories, science and epistemologies. But while some human communities indeed live with the sea on close terms (McNiven 2016), and our bodies largely consist of water (Neimanis 2017), other animals and plants live in the sea, and for that reason alone already have a unique perspective on the water. They also depend on the sea, especially when it is their home. When considering how the Dutch can learn to think and act with, instead of just about, or against, the sea, they are the first beings that should be consulted.

The North Sea is a home to many sea and coastal animals, like sea anemones, sponges, crabs, lobsters, shrimps, dolphins, seals, sharks, shellfish, plankton, whales, squid, gannets, gulls, and many others. Between humans and other animals there are many encounters, in the sea and coastal areas. Most of the encounters that Pika and I had on the beach were with sea birds like herring gulls and sandpipers; we also saw stranded jellyfish and starfish, once a flatfish, whom I returned to the sea, and once a seal who was sunbathing. On the collective level, most encounters between humans and sea animals are violent and take place in the context of fishing (Wadiwel 2016). The Dutch sea fishing fleet consists of around six hundred ships (Noordzeeloket),[4] which are divided into coastal and North Sea fisheries, large sea fisheries (pelagic freezer trawlers), shellfish fisheries and gillnet fisheries (Noordzeeloket, CBS).[5] The main species that are being hunted are sole, plaice, langoustines, shrimps, mussels, and oysters (Noordzeeloket, CBS). Policies, for example with regard to fishing zones and species that can be fished, fall under European legislation; the first three kilometers off the coast are reserved for Dutch fisheries (Noordzeeloket, CBS). There are also less harmful encounters between humans and fishes, and other sea beings, like shrimp or seals. Some forms of research are benevolent (Balcombe 2016); underwater photography and diving can be too. For animals in need, we have the animal ambulance, and for seals who need assistance there exists a seal ambulance; there are wildlife rehabilitation centers for seals and other sea animals on the Dutch coast and the island of Texel, like Ecomare.

While the kinds of encounters and interactions between humans and sea animals vary, they are generally are fleeting and more similar to the interactions with the amphibians I discussed in Chapter 4 than to the intimate conversation I had with Olli. Humans and sea animals in the Netherlands do not form shared communities.[6] Most, if not all, sea animals prefer to live in their own communities and do not need human assistance in order to flourish. When humans and sea animals do enter into a conversation, for example in the context of research or rehabilitation, their interactions share key characteristics with the multispecies conversations I have discussed so far: they involve more senses than only sound and sight (for example, touch can play a role in rehabilitation processes), they are embodied and

4 https://www.noordzeeloket.nl/functies-gebruik/visserij/. Accessed October 1, 2024.

5 Precise numbers and more information can be found at the website of the CBS: https://www.cbs.nl/nl-nl/onze-diensten/methoden/onderzoeksomschrijvingen/korte-onderzoeksomschrijvingen/visserij-vanaf-1930. Accessed October 1, 2024.

6 Sometimes humans and sea animals do form collectives. For example, fishermen and dolphins fish together in Laguna, Brazil (Peterson et al. 2008).

tied to the specific context in which they take place, they often include objects (boats, fishing nets, diving equipment), and they are shaped by the agency of the different beings involved. Some of these interactions, as in the context of research of large sea mammals or assisting animals in need, may closely resemble human language-games that we call conversations; they involve questions and responses, and they can be improved if humans become more attentive. Other interactions may be very different.

Abolishing violence and becoming more attentive should be central in refiguring the conversations we have with sea animals. Many sea animals, like fishes, are currently not seen as subjects by humans and they are not consulted with regard as to how we share the sea (Balcombe 2016, 2022). Recognizing the subjectivity of these animals, and viewing interactions as conversations or deliberation, would challenge the foundations of the fishing industry and related industries, so change is opposed by economic forces (see Wadiwel 2015, 2023). But the problem is not just economic, it is cultural too (Balcombe 2016; Braithwaite 2010; Wadiwel 2016). Sea animals are seen and treated as objects, not as inhabitants of the sea who have a right to live there. This bias also long affected scientific research and knowledge production about them. For example, because humans doubted if fishes feel pain, there was little research into that topic and it took until 2005 before fish biologist Victoria Braithwaite wrote the authoritative book on fish pain. We now also know they can use tools, make friends, work together, and communicate with each other and members of other species in complex ways (Balcombe 2016). Human acts not only harm them physically but also emotionally: research shows that salmon become depressed and even suicidal in the fishing industry (Vindas et al. 2016). Not taking fishes seriously and treating them violently is problematic for reasons of justice – these animals are sentient beings with interests, whose homes are invaded and communities are torn apart through human activities like fishing, sound pollution, and other forms of pollution – but it is also unfortunate from the perspective of learning to live better with the sea. Humans who live in close relation with the sea, such as fisherfolk or beach lifeguards, often become adept at reading it – they have to. But sea creatures have an unparalleled access to, and perspective on, the saltwater world. They also know how to relate to it sustainably, which matters when thinking about the future.

Speaking with seaweed

Dutch writer and artist Miek Zwamborn writes that the same is true for seaweed, and, even stronger, that seaweed holds the key to more sustainable relations in general. 'They can teach us how to live,' she writes about seaweed species. '(H)ow to form strong communities, how to learn to become more flexible as a species and also how to nourish ourselves in a healthy way' (2022). Zwamborn has written extensively about seaweed (2020, 2022) and her descriptions often resemble dialogues between her and the algae. Her work not only gives an impression of the kind of shape that conversations with plants can take, but also shows an alternative to human supremacy in dealing with plants.

Zwamborn is a visual artist, poet and novelist who moved to the Isle of Mull together with fellow artist Rutger Emmelkamp in 2017, to take care of a nature reserve called the Tireragan Estate. They founded an artists' residency called Knockvologan, where they aim to develop a new way of living with the landscape and nature. One of the ways in which Zwamborn does this is by focusing on seaweed. She writes about seaweed (2020, 2022), but also draws seaweed, cooks with the plants, uses them as a fertilizer for her garden, and experiments with other uses (2020). In her work, there is a strong recognition of seaweed agency. In fact, it was the seaweed that began this human-plant dialogue:

> The mantle of seaweed that encircles Mull has been directing me from the day I arrived five years ago. Walking to the Scottish peninsula of Erraid at ebb tide, I came across an immense chunk of kelp beached at the tidal flat. Its dark leaves and stems contrasted strongly with the pale sand. The seaweed looked lively and spiraled like an Archimedean screw. [...] The seaweed plant seemed to be filled with a strange force. It appeared boundless. This deep encounter at the bottom of the sea was like a call, an invitation to another universe that I didn't know anything about. In the year that followed I collected seaweed in all sorts: green, red and brown varieties and every color that lies in between – spotted, perforated, translucent, albino. I cut blades and receptacles of seaweed from the rocks, plucked them from the surf or picked them up along the tideline. [...] This was how it began. I had found something to hold fast to. (Zwamborn 2022).

Through swimming through, drawing, writing, and eating seaweed, Zwamborn investigates the boundaries between human and seaweed. According to her, these are less strict than is usually assumed. Writing about seaweed asks for giving up one's sovereignty: 'As an underwater writer you have to discard everything, be submissive, strap on lead blocks; otherwise, you will float back up to the surface and all the newly written words will ebb away' (2020, 58). Looking at seaweed invites thoughts of becoming like seaweed, as Zwamborn describes in relation to a kelp leaf to which a brocade of Obelia geniculate is attached: 'All these travelling particles that carry life within them and will settle on the seabed or some other surface have a strange effect on me. Will they settle on me too? Is the water making me permeable?' (2020, 59) Becoming seaweed offers a vision for living differently with others too: 'In all this interconnectedness, you might see an example of an ideal world in which species are tolerant and offer each other holdfast in the current in order to survive' (2020, 59).

In addition to describing direct encounters with seaweed, in which embodied and tactile entanglements are foregrounded, Zwamborn maps cultural practices around seaweed. For example, she describes the role of seaweed in the myths, fairy tales, and folktales of many geographical regions and areas, showing how these plants leave their mark on human communities (2020). She also draws attention to the role that seaweed can play in working towards a more sustainable future, specifically in relation to eating, design, and farming practices, and therefore in addressing the climate crisis. Seaweed produces oxygen, around 70% of the oxygen on earth, and it absorbs carbon – kelp for example absorbs five times more carbon than land-based plants (Zwamborn 2020, 118). Seaweed farming is seen as ecologically sustainable, making it a renewable and healthy food source, which could be of great importance in addressing the world food problem (Tiwari and Troy 2015). It can also be used for making cosmetics, fertilizer, fuel (research is being done into using seaweed as a sustainable alternative to kerosine for airplanes), medicine, agar, and many other products (Tiwari and Troy 2015). Zwamborn emphasizes that transitioning to farming seaweed on a greater scale also requires that we should work towards a more sustainable relation with seaweed because currently pollution and global heating are threatening the health of these plants. Furthermore, while seaweed farming does not necessarily kill the plants, one should not take too much. Zwamborn recommends taking not more than 30%.

In describing conversations with seaweed, Zwamborn lays out an indirect model of question and response. Seaweed affects her thoughts and feelings through tactile encounters in the water and on the land, and through these encounters her behavior towards seaweed changes. She works with seaweed carefully, and translates her insights to fellow humans, through text and visual art. The aesthetic and ethical dimensions of this project are intertwined: awe for seaweed can inspire a different attitude in humans, which is a starting point for learning to live differently (Blissett 2021).

The way that Zwamborn writes about seaweed shares similarities to the conversations with animals I discussed above and in earlier chapters, but there are also significant differences. While Zwamborn gets to know the seaweed, it is difficult to speak of an intersubjective relationship. Seaweed exercises agency, but in other ways than more-than-human animals do, and the understanding does not grow over time, as, for example, in the case of the mice in Chapter 3. This has consequences for the kind of conversations that are possible with these different beings. Humans have responsibilities towards seaweed, but these differ from responsibilities towards nonhuman animals (Gaard 2016). The normative significance of having conversations with plants is furthermore less clear than with nonhuman animals – with animals, conversations have a clear function in interspecies deliberation (Meijer 2019), but for improving relations with seaweed, deliberation might not hold the same importance. In order to be able to conceptualize what is at stake in the conversations between humans and plants on the one side, and humans and animals on the other, in relation to speaking about the sea, I first want to take a closer look at differences and similarities in critical scholarship about animals and plants.

Mapping relations between critical plant studies and critical animal studies

The fact that the human was seen as the standard in biology and the humanities for centuries has influenced our knowledge of other beings (Gaard 2016; Lawrence 2022; Probyn-Rapsey 2018). For example, nonhuman animal and plant intelligence were long studied on the basis of how much other animals and plants resembled humans, which led to a distorted view of their capacities, and often their moral worth too (Gaard 2016; Lawrence 2022; Probyn-Rapsey 2018). This human bias is increasingly criticized (Gaard 2016; Lawrence 2022; Probyn-Rapsey 2018), which also affects public discourse about animal capacities, their inner and social lives (i.e. Bekoff 2012; De Waal 2016), as well as plant capacities and social lives (i.e. Mancuso 2018; Wohlleben 2016). Following this scholarship in the life sciences and public discourse, scholars in the humanities are rethinking relations with more-than-human beings, notably in critical animal studies and critical plant studies. These are umbrella terms for academics working in different disciplines, ranging from philosophy and sociology to geography, who connect insights about agency, subjectivity, and related themes to a critical investigation of the power structures that govern human relations with animals and plants respectively. While there are many differences between plants and animals, and their respective relations with humans, beings who are seen as part of the categories 'animals' and 'plants' are similarly affected by human bias in academia and society: their oppression shares characteristics under the conditions of anthropocentrism. Furthermore, many ethical and political questions, and systems of power more generally, concern both nonhuman animals and plants. However, in analyses of power as well as theories of justice, plants and animals are usually discussed separately.

Critical plant scholars generally focus on 'humans' and 'plants', and do not include nonhuman animals in their analysis, although other animals also suffer from human violence, and are often part of human-plants entanglements too (see Gaard 2016 for a longer discussion of this phenomenon). An example of this is found in the work of Zwamborn too. In a discussion of the benefits of seaweed for sustainability, she mentions that growing seaweed to feed farm animals needs neither water nor fertilizer, and would free up land that can be used for other crops and fruits for human food, without problematizing the instrumental use of nonhuman animals in farming. She also mentions climate benefits in relation to animal use: it reduces methane output (2020, 116), without mentioning the violence inherent in using animals for food. In this understanding, benefits of using seaweed are benefits for humans, not for all animals. It is theoretically contradictory to stress interconnections with plants and search for better ways of existing with them as the basis of a sustainable and holistic attitude, while neglecting human exceptionalism and violence towards other animals, but it is also morally problematic because it keeps intact large-scale violence towards nonhuman animals.

Similarly, critical animal studies scholars do not often take questions of plant ethics seriously, often because they see 'sentience' as the threshold for moral consideration and plants were long not considered to be sentient (see for example Cochrane 2018). Sentience is, however, a problematic concept, the scope of which continually changes under the influence of research and shifting power relations (Duncan 2006), which in recent times has changed with regard to insects such as bees (Buchmann 2023).[7] New research leads some scholars to include plants (Mediano and Calvo 2021; Segundo-Ortin and Calvo 2023). Scholars who focus on those animals who have a self, or in whom there is 'someone home' as Donaldson and Kymlicka describe (2011, 25), encounter a similar problem, because it is unclear what counts as a self, and who has the power to decide that (Oliver 2009). More importantly however, rethinking human exceptionalism implies considering the whole web of relations of which humans are a part, and in which humans and other animals depend on other living beings. Doing justice to multiple actors and

7 See also the New York Declaration of Animal Consciousness: https://sites.google.com/nyu.edu/nydeclaration.

the web of relations they are part of asks for a holistic perspective which is attentive to the flourishing (and overlapping oppressions) of different beings.

Vegan ecofeminism: Connections between thinking about plants and nonhuman animals

One branch of thought that does consider plants and animals in tandem is vegan ecofeminism. In vegan ecofeminism, connections are made between the oppressions of women, nature, and animals. Greta Gaard (2016; see also Donovan and Adams 2007) draws attention to the importance of thinking about plants for vegan ecofeminism, and proposes to view CPS and CAS as connected fields. As a starting point for reframing human relations to plants and nonhuman animals, Gaard maps similarities and differences between the groups. Both plants and nonhuman animals have senses, intelligence, forms of communication, self-identity, and unique features that give rise to moral standing and/or rights. There are also many differences between these two groups, for example with regard to what specific senses beings have, and what forms of communication, intelligence, and identity. She adds that it is worth noting that there is no clear line between the categories 'plant' and 'animal', and that the groups themselves are heterogeneous. Humans and octopuses have very different ways of experiencing the world and communicating; oak trees and algae too.

While Gaard investigates plant and animal capacities, she also problematizes human categorizing. Human bias is interconnected with backgrounding animal and plant agency and material exploitation of nonhumans (Lawrence 2022). As the examples of the mice and the fishes show, knowledge production is dynamically related to larger structures of normativity in society, and if academics are not critical of power relations they generally tend to repeat them (Despret 2016). In order to avoid this, Gaard (2016) proposes to rethink what she calls 'main categories' of meat, species, and plant, that govern our understanding of fellow beings. Instead of defining similarities or differences from a human-centered perspective, and affirming existing categories, Gaard proposes to focus on relations. She writes that humans, other animals, and plants are connected through webs of interdepence in which different beings occupy different positions of power, have different interests, and different ways of flourishing. Developing ethical relations means something different with regard to plants than with regard to animals. An example is eating others. Animals are killed to be eaten, and also for other 'products' (Adams 1990) such as their milk or eggs. Plants, however, can continue to live after losing 90% of their bodies. This has normative consequences for living justly with these beings and to respect their flourishing. All systems of food production have harmful side effects, which might include the deaths of, or harm to, plants and animals (including humans), but this does not mean that they are all harmful to the same degree.

Rethinking categories as a basis for thinking and acting differently should include a critique of ideological systems, such as anthropocentrism. What is considered to be natural is generally a cultural construction, and Gaard explicitly draws attention to how our eating practices and academic thinking are shaped by capitalism and heteronormativity (Mortimer-Sandilands and Erickson 2010). In order to develop a 'contextual moral veganism' (2016, 276) we need to undo 'the grasp and hegemony of a carnist culture' (2016, 280). This requires 'shifting from denial to attentive listening, from alienation to empathy, from capitalist production time to seasonal time, from a heteronormative universalism to a queer multiversalism' (2016, 280). Making this movement asks for a change in perspective which is rooted in embodiment not consciousness, and for which Buddhist and other non-western philosophies can provide inspiration. In order to overcome carnist patriarchal systems, we need to learn to take the perspectives of plants and other animals (Plumwood 2014; Meijer 2019).

Taking into account the perspectives of more-than-human beings is important as an ethical movement, in individual relations, and as a cultural position. But it is also a political project. Gaard (2016) stresses the politics of vegan ecofeminism, and offers alternatives to hegemonic power struggles. These alternatives ask for attention and listening, to better understand co-being, and for recognizing the multiplicity of the relations we are part of.

Speaking with plants and animals about the sea: silencing and backgrounding

The main lessons from vegan ecofeminism for thinking about conversations with plants and animals are that it is important to consider relations with both entities in tandem, and that there is much we do not know about their capacities and possibilities for interaction because of power structures that limit their behavior and our knowledge. Both plants and nonhuman animals are silenced, their agency is backgrounded, and this ongoing silencing is part of their oppression (see Meijer 2022d). We do not know much about the conversations that currently take place around us because we are not interested, and we also do not know much about improving conversations because we lack the knowledge and understanding. Changing this does not ask for more research into their capacities in a knowledge system aimed at use, but for respectful engagement on their terms and a critical investigation into patterns of backgrounding and silencing.

In relation to seaweed and plants more generally, the work of Zwamborn provides inspiration. She challenges the neglect of seaweed in relation to human superiority by foregrounding seaweed. Seaweed is often literally invisible, though sometimes it is revealed by the sea at low tide, which makes it easy for humans to disregard it. But this disregard is also present symbolically: humans do not see seaweed for what it is (Zwamborn 2020). This is true of other water and land plants too, leading critical plant scholar Catriona Sandilands to speak of 'systemic insensitivity' towards plants in many cultures (2021). Not all humans disregard plants though; in many indigenous communities plants do play a central role, both in practices and epistemologies. Robin Wall Kimmerer (2013), for example, writes that plants are seen as teachers in different indigenous traditions of thought: because they have been on this planet longer than humans, they know more about life than they do (2013, 9; see also Sandilands 2021). But in western philosophy and many societies worldwide, plants are neglected (Lawrence 2022; TallBear 2011).

Ecofeminist Val Plumwood (2001) calls the invisibilization of plants, and nature more generally, 'backgrounding'. Backgrounding takes place in frameworks of domination, and involves a simultaneous reliance on and disavowal of the agency of subordinated others. When the dominating party sees themself as radically different and separated from others, they often also devalue the agency of these others and make it invisible. An example that Plumwood (2001) mentions is unpaid labor of women, including house labor and child raising, which in androcentric systems is backgrounded, even though it makes other forms of labor (that are regarded as important) possible. The agency of nature is similarly taken for granted and dismissed, and Plumwood sees the backgrounding and othering of nature as the most problematic side effect of human's attitude towards the nonhuman world. Challenging it is therefore an important part of reconceptualizing the ethical position of humans. Backgrounding also takes place in relation to nonhuman animals. For example, Wadiwel (2015, 2016) shows that fish agency is systemically obscured in western societies, which masks violence towards them.

In the context of language, the backgrounding of animals and plants is reinforced by silencing. Their languages are not seen as language, and this is presented as an empirical issue while it is, in fact, the outcome of a system of domination. Furthermore, this also strengthens domination because their perspectives and voices are not taken seriously democratically (Meijer

2023d).[8] Speaking better with sea animals and plants therefore begins with acknowledging that they have something to say, and challenging human supremacy. What animals and plants have to tell us about the sea – and what they want to tell us – will only become clear once we begin to listen, and practice attention.

Speaking with the sea

Listening also matters with regard to getting to know the sea as a whole. Painter Betzy Akersloot Berg (1850–1922) devoted her life to painting the sea, often working on the beach of Dutch Wadden Island Vlieland, where she lived from 1896 until her death (Haselmann and Klok 2021). Akersloot Berg was born in Norway and trained to be a nurse, but she wanted to travel and paint. She took lessons at the drawing academy in Oslo, and followed her teacher, painter Otto Sinding, to Munich where she became acquainted with the work of Hendrik Willem Mesdag, a Dutch maritime painter. She lived in Paris and Belgium before settling on Vlieland with her husband. Her studio was located behind their house (which is now a museum), overlooking the sea, but Akersloot Berg also often painted outside, on land and on ships. She was particularly fond of bad weather, and wore an oil jacket while painting out on the beach, sitting in an open chest made especially for this purpose (Haselmann and Klok 2021). Because she liked painting under harsh weather conditions, she also witnessed shipwreckings.[9] Through painting, Akersloot Berg engaged with the sea and fellow humans – the latter conversation is ongoing, because her paintings are still on view in Tromp's Huys in Vlieland. Akersloot Berg is buried at the cemetery on Vlieland, and under her name is written 'sea painter'.

Similar to Zwamborn, Akersloot Berg took time to get to know a nonhuman agent, and used artistic methods to engage in interaction. The sea spoke to her visually (and through other senses) and she responded in the same manner. During her years on Vlieland, her style changed, the strokes

8 For this reason, it is problematic to enter into discussions about whether or not nonhuman animals have consciousness, language, culture and other capacities and forms of organization that were long seen as the territory of humans. Entering these discussions legitimates the doubt – as if it were up to humans to determine whether others speak, feel, etc. – and keeps intact a framework in which nonhumans are systematically silenced and backgrounded.

9 https://www.femartcollection.org/betzy-akersloot-berg/. Accessed October 1, 2024. And: https://resources.huygens.knaw.nl/vrouwenlexicon/lemmata/data/berg. Accessed October 1, 2024.

became bolder and her use of color brighter. Maybe this happened through the influence of Mesdag or the rise of impressionism, or maybe it was the sea itself that caused this change. If you look at something for a long period of time it becomes part of you, and changes you.

By working and living in this way, Akersloot Berg enacts what Jane Bennett calls a vital materialist approach to the sea (2015, xvi). Vital materialism is a materialism that recognizes the forces at work in nature, human bodies, and human artifacts, in which 'things formerly known as objects' (2015, xvi) possess a liveliness that invokes feelings and acts in humans. In vital materialism, agency is understood as a dynamic force and not a capacity of the subject (or object), which is always found in assemblages, coming-togethers of human and nonhuman forces (Bennett 2015, Chapter 2; see also Latour 2007, 2017; Tsing 2015).

This understanding of agency, and the dynamic interconnections between different kinds of beings in assemblages, allows us to understand how Akersloot Berg is formed by painting the sea and how interactions between very different kinds of beings constitute her paintings. Akersloot Berg is part of an assemblage in which the sea, the beach, other humans on ships, ships, sea birds, paint, canvas, salt, air, and other factors all co-determine the outcome. While speaking of sea agency is contested in philosophy (Meijer and Bovenkerk 2021), acknowledging the influence of nature on human bodies, minds, and feelings does connect to common sense understandings of the sea. For example, many humans report feeling better after taking a walk along the beach, and those who live near the sea might miss its presence when they spend too much time in a city.

The sea shapes not only the days and feelings of those individuals who commit to painting the sea, but also communities and cultures. For example, living in close proximity to the North Sea and Wadden Sea has historically shaped the economic practices and cultural self-understanding of the Dutch. Historian Michael Pye (2014) shows how the North Sea changed and influenced the lives of its coastal peoples. He describes how medieval communities gradually became more reliant on the sea, not only using it for fishing but also traveling its surface to discover new lands, go on pilgrimages, abduct women, and enslave other humans. The sea gave rise to the Frisians and later the Vikings, and was the source of their prosperity. The Vikings were so proud of their seafaring way of life that they buried their dead in stone ships. The sea provided new forms of wealth and created the conditions for knowledge transfer, but also took lives – for example, when it conveyed the bubonic plague to Europe. Pye tells us the sea also brought forth new forms of bureaucracy, because different countries needed a uniform bookkeeping system for overseas trade. This system was set out in laws and other documents; words, too, can travel the seas. In short, the North Sea, Pye argues, was partly responsible for the ultimate form of the societies around it.[10]

It is important to note that in sea-assemblages, different types of agency (Meijer and Bovenkerk 2021; Knudsen 2023) including different structural forces – economic, social, political, biological, and others – are at play. In order to understand how we are co-constituted by nonhuman forces, we do not have to subscribe to a flat ontology, in which all forms of agency weigh the same, and from which it is difficult to formulate ethical and political demands and duties (Meijer and Bovenkerk 2021; Knudsen 2023). The various nonhuman animals and plants that exist express different forms of intentionality and agency, and humans and the sea do too, and all of us are affected by many types of acts from other agents and forces (ranging from economic human structures to the wind). This too connects to common sense, and is expressed in everyday language, for example, in how the concept 'agency' is used by non-philosophers. It is common to say that the weather acts in a certain way, or that animals act. There are different forms of agency. As Wittgenstein writes, the philosophical desire to find one universal meaning of concept obscures how language works, as well as the phenomena that are referred to (1958). Defining 'agency' in one way as humans, furthermore runs into the problem that humans are tied to their own position in the scheme of power and knowledge and have historically privileged human modes of agency (Pearson 2017; see also Derrida 2008 on *logos* and animals).

Acknowledging the relational aspects of agency as well as differences in agency is important for evaluating whether or not conversations take place, and what they entail. While it makes sense to say that Akersloot Berg is in conversation with the sea, it is also clear that 'conversation' in this case means something other than it does between a human and a dog. Before I discuss why I still think it matters to speak of conversations, I want to discuss two cases in which humans look for new relations with the North Sea: The Embassy of the North Sea and the Sand Motor.

10 The influence of the sea on humans can go further than the social and cultural, to the level of physiology. The Bajau, a nomadic people who live by fishing in the waters of the Philippines, can remain underwater for up to thirteen minutes without breathing; most people can barely hold their breath for two minutes. Like whales, they have enlarged spleens. Researchers believe that they evolved in this way by living with the sea (Gibbens 2018). The bodies of many other sea creatures – whether they have wings, scales, or shells – have been shaped in similar ways as that of the Bajau by conditions in or on the sea.

Listening to and representing the sea

The Embassy of the North Sea (*Ambassade van de Noordzee*) is a Dutch long-term initiative in which artists and scholars think about and experiment with political relations with the sea. Their aim is to listen to, speak with (and imagine), and represent the North Sea as a full political agent. The core team works together with many different artists, scholars, and organizations. They organize lectures, art exhibitions, workshops, and other sub-projects to investigate how humans can better listen to the sea, speak about the sea, speak with the sea, and represent the sea politically. While some of the projects are theoretical in nature, for example those that focus more strongly on representation in law and take a written form, others actively engage with the sea and water beings, for example through organizing expeditions to get to know the dunes or inviting humans to smell the scents of the sea. Language plays an important role in the project, and the organizers explicitly mention listening to the sea, and speaking with the sea. In their press releases they often use the phrase 'in conversation with'[11] and one of the subprojects searches for new concepts and words to be able to speak with the sea.[12] Through the use of artistic and scientific methods, the Embassy of the North Sea does not have one conversation with the sea, but sets up several, between different human parties, coastal and sea beings, and the sea itself.

Speaking through sand

Another collective conversation with the sea takes place in Kijkduin, near The Hague, where humans constructed an artificial sand bank just off the coast in 2021, called the Sand Motor, for environmental and scientific reasons (Sand Motor). For this new peninsula, 21.5 million cubic meters of sand were used, and it was designed following the principles of 'Building with Nature'. As the project's website explains: 'The Sand Motor is a unique experiment, because it works *with* water, instead of against it. Depositing a large amount of sand in one go prevents the repeated disturbance of the seabed. Nature then distributes the sand to the right places for us' (Sand Motor). The Sand Motor is described as a tool to combat climate change and an experiment in coastal protection; its aim is to make the coast line stronger for the future. This experiment is set up as a conversation with nonhumans: humans offer sand, while the sea, waves, wind, tides, plants, and animals respond with

11 For example, with the dunes: https://www.ambassadevandenoordzee.nl/events/gastcollege-in-gesprek-met-de-duinen/. Accessed October 1, 2024.

12 https://www.taalvoordetoekomst.nl. Accessed October 1, 2024.

their acts and movements. New animal species, including shorebirds, and plants settled in this artificial-natural landscape, and a lake and laguna formed, that are changing shape and might disappear later on. After five years, dunes began to form; they need sediment and vegetation, and it took some time for the sand to arrive in the right spot because of the lake and the laguna that were formed by the tides and weather. Different human parties are involved in the project – Rijkswaterstaat, the Dutch water guards, the province, scientists and designers – and they have different aims, such as gaining knowledge about coastal protection, increasing coastal safety, providing humans with space for leisure, and creating opportunities for nonhuman beings to flourish.

Stefan Helmreich (2023) sees the Sand Motor as a typical example of the Dutch attitude towards nature and the sea. He describes the experiment as a form of domestication, of control, which uses waves and wind, and as a model that can be transported to other places in the world. While I agree with Helmreich's analysis of the Dutch culture of control and cultivation, the situation changes if we view the Sand Motor as a starting point for dialogue. While humans started the experiment, the North Sea co-determined the course of the project and its outcomes. Humans made models, but were not sure how and if the real situation would turn out, and unexpected events occurred, such as the slow development of the dunes, and visits from new animals. The Sand Motor shows that this type of intervention in the landscape can function as a starting point for dialogue, not only between humans and the sea or the landscape more generally, but also with other animals and plants, entities that Helmreich does not consider.

Material conversations

In the two examples above, humans try to establish a relationship with the sea and/or its inhabitants over time, which is characterized by receptivity on the side of the humans. The Embassy of the North Sea and the Sand Motor explicitly recognize nonhuman agency as the basis for interaction. With regard to the relation between Akersloot Berg and the sea, I mentioned that everyday language allows us to call it a conversation, but that it is immediately clear that this conversation is different from that between two animals. It unfolds visually, and relies on the determination of Akersloot Berg. Similarly, the Sand Motor speaks with the sea through adding sand, and while the sea reacts, it is unclear whether it *responds* (Derrida 2008).

But maybe asking whether the sea responds is the wrong question, and instead we should simply accept that the sea speaks in a different way than humans do (McNiven 2003, 2016). A skeptic could here object that we do not need to use the word 'speaking' and can rely on 'acting' instead. However, as I discussed in the introduction, the figure of the dialogue is suitable for rethinking intersubjectivity, and in this case using the concept 'dialogue' is also a normative intervention. Recognizing natural entities, more-than-human animals, plants, and ecosystems as interlocutors is a gesture of respect that can guide more respectful relations (McNiven 2003, 2016; Kimmerer 2013). Shifting the focus from 'Is this a conversation?' to 'How can we have a better conversation?' is fruitful because it invites us to overcome (often biased) human views of others. This does not mean throwing existing ideas about the concepts 'language' or 'conversation' overboard – these are the stepping stones we need to make sense of what we experience. Nor will viewing others differently lead to harmony or a full understanding of them – it will likely add to the already existing complexity of living well with others. But it does change the question, and in the context of large-scale environmental degradation, industrialized animal exploitation, global warming, pollution, and the extinction of species, that is a necessary movement.

Becoming better interlocutors can be understood as a virtue ethical project. Similar to how we encounter humans, as individuals we can treat the sea, plants, and nonhuman animals as if they have something to say to us. We can spend time on the beach, in the dunes, swimming in the water, learning about the coast, speaking with other animals (or accepting their refusal), and so on. Through respectful interaction we can learn to have better dialogues with these beings. But there is also a more political way of approaching the normative dimensions of our relation with the sea. As communities we have long seen natural entities as mute, as raw material, or as force to be conquered. In the Netherlands, the latter discourse still prevails: the Dutch are proud of their water management, and feel entitled to life in the sea (Noordzeeloket; Pye 2014). Seeing the sea as interlocutor changes social and political conversations, and can help humans to ask different questions. Education, culture, and art have a role to play here, but political practices and institutions do too. This leads us to the question of deliberation with the sea.

Deliberation with the sea

Similar to dialogues, deliberation with the sea will look different from deliberation with humans. In order to foster more-than-human deliberation, we need to map existing interactions and investigate where more-than-human agency is silenced, ignored, or glossed over; set up political experiments; and extend existing institutions to also acknowledge the interests of nonhumans.

The first step in conceptualizing human-sea deliberation is to investigate which interaction already takes place, what kind of questions and responses are at work, and how sea agency can be brought to the front. In the Netherlands, deliberation by humans about the sea now mostly takes place in the context of the climate crisis and concerns human safety (NRC 2022). Different human parties take part in this deliberation, amongst whom are scientists, politicians, activists, and BN-ers (famous Dutch people), who mostly express their opinions on X and Bluesky, and in talk shows. Animal voices are absent, and other nonhuman voices too. As we saw, however, there are different initiatives that focus on working with nature to create a stronger coastal line, like the Sand Motor, and there are similar initiatives working at changing the landscape further inland, for example by giving natural areas back to the sea (NIOZ 2022). Mapping how different agents act and speak in these processes, and mapping the relations between these and human forms of deliberation (i.e. investigating how human deliberation is shaped by more-than-human agency), can help inform political deliberation (see also Hobson 2013). As with nonhuman animals (see Meijer 2019, Chapter 9), this requires attending to processes instead of immediately making decisions, changing where we deliberate – i.e. move the focus from official political settings to dunes and beaches – as well as attending to the embodied and situated character of deliberation (Dryzek 2000; George and Manzo 2022) and its material dimensions. We need to become attentive to aquatic agency (McNiven 2003; Helmreich 2023).

The second step in developing new practices of deliberation is experiment. As the Sand Motor and the different projects of the Embassy of the North Sea show, there is much that we do not know yet about the sea, its inhabitants, our capacity for engaging with nonhuman actors, and their capacity (and desire) to engage with humans. Through material interventions we can ask the sea questions, and through attending to the sea carefully, as Akersloot Berg did, we can learn to see it and ourselves differently. Further exploring what dialogues could entail and what they can bring us requires small-scale experiments in coastal areas, focusing either on the sea itself or animals and plants who live in or near the sea. The Sand Motor is a scientific experiment, and the Embassy of the North Sea uses art and science to rethink relations, but from the perspective of deliberation it is important to investigate the political aspects of these projects. This has a conceptual side, for example with regard to investigating what democracy or justice means in sea-assemblages, but it also means bringing politicians and/or citizen's assemblies in conversation with artists and scientists. Again, this project is not solely human: experiments can also help to shed light on how to better listen to, and include, animal and plant voices.

This connects to the last point that I want to make about sea deliberation: in order for the voices of sea beings and the sea itself to be heard and listened to, existing political institutions and practices should be extended to include them. For example, rights can be extended to also protect seas and their inhabitants, and within official forms of politics, such as parliamentary politics, the sea and/or its inhabitants can be represented formally (Tanasescu 2016). As I will discuss in more detail in the last chapter, new listening practices can inform representation and lead to new habits and rituals, and more generally another attitude towards nonhumans (Meijer 2023a). Education has a role to play here too: in learning about the sea, and how to co-exist with the natural world more generally, both with regard to children and adult humans (Chapter 8).

Importantly, deliberation does not mean more intervention. Humans need to take responsibility for their acts – in the context of the sea for example for ending chemical and sound pollution, and of course ending the killing of fishes and other sea beings. An important aspect of this is retreating as humans. Treating an entity as an interlocutor also implies a respectful attitude, including respecting their agency and dignity. In the Netherlands, nature reserves on land are currently dominated and managed by humans (Keulartz et al. 2004), which affects the agency of the animals and plants who live there, and often leads to violence against them in the name of management, for example through hunting. In the nature reserves on sea, humans too have the last word, in deciding who is safe where. One way of countering this is by installing 'no-entry zones', which, for example, benefits sharks (Robbins et al. 2006). This would not only provide the animals and plants who live in these areas with sanctuary, but also gives them a stronger position to speak from.

In non-western traditions, alternatives to human-centered models of sharing habitats already exist (Biemann and Tavares 2014; Kimmerer 2013; Simpson 2017). For example, McNiven (2003, 2016) discusses relations between Saltwater Peoples (the marine specialists among Aboriginal people and Torres Strait Islanders) and the sea through an analysis of seascapes, which not only offer resources, but are spiritscapes too: home to spiritual entities, which play an important part in the cosmologies of Saltwater Peoples. In these cosmologies, ancestor spirits 'imbue seas with spiritual energies, fecundity and sentience' (2003, 332) and different features of the sea such as tides or island, as well as sea creatures, are created by these ancestors, in a mythical past (2003, 332). The sea is seen as a giver and taker of life, which in its dynamic changes (of colors, temperature, waves) and agency resembles the sentient spiritual beings who animate it. Interaction with the sea is sometimes described in linguistic terms. McNiven describes that for the Gidjingali of Arnhem Land 'seawater and waves are imbued with spiritual essences that are expressed anthropomorphically as "habitual speaking" (beach surf) and "habitual walking" (currents)' (Bagshaw 1998, 159, as cited in in McNiven 2003, 333). Active engagement with the seascape as spiritscape encompasses many practices, rituals, and interventions in the coastal landscape that can be seen as a material dialogue between humans, ancestors, and the sea, in which the movements of the sea co-creates meaning. In the history of Europe there is also a tradition of viewing nature differently too, which includes nature spirits and gods, as the example of Nehalennia that I discussed in the beginning of this chapter shows (Wagenvoort 1971). Recovering this tradition can perhaps also provide guidance.

While different cultural traditions and historical practices can help reframe conversations with the sea, developing a blueprint for deliberation with the sea is impossible without engaging differently with it.[13] Working towards more responsive and responsible relations with the sea will not change the sea, but it will change humans, which is needed for building new worlds with others. In this process, rethinking concepts like 'dialogue' or 'deliberation' in relation to new entities does not mean they have to fit into a certain mold. Rather, it is an open-ended inquiry into what concepts mean and can mean. I therefore want to end this chapter with a brief discussion of the status of concepts in conceptualizing relations and conversations with others, and the need for new narratives.

The role of language in moving forward: rethinking concepts, and devising new narratives

Developing a multispecies multidimensional model of conversation asks us not only to pay attention to linguistic interaction, but also to how human language works. Concepts may have different meanings in different situations, but this we often do not notice. '(T)he clothing of our language makes everything alike,' Wittgenstein writes (2010, 224). Words deceive us because

13 In Iceland, the glacier Snæfellsjökul entered the presidential elections of 2024, supported by a large team of artists and researchers. In speaking with and about the glacier, the humans in the team use insights from different traditions. For example, the knowledge about the *huldufólk* (hidden folk, sometimes also called elves) that exists in the Icelandic culture, but also cultural knowledge for engaging with natural entities from Mexico and Mongolia, western scientific knowledge about nature, animals and plants, and artistic experimentation (personal correspondence, see also: https://www.kjosumjokul.com/. Accessed October 1, 2024).

they have the same shape in different situations, which makes it seem as if they always mean the same thing, while in fact their meaning varies depending on the different language-games in which they are used. Words have histories, they change over time and in different settings. In order to understand what concepts like 'conversation' or 'democracy' exactly mean, and envision what they could mean, we need to critically examine them in view of the power relations that gave them their shape, and recognize that language is always situated and embedded in socio-cultural practices. There is no ahistorical universal meaning for any word or concept; instead there are many language-games in which concepts gain their precise shape. The meaning of 'rights' for example has changed under the influence of social movements, the declaration of universal human rights, the rise of the nation state, struggles for women's rights, and many other historical events and processes.

Mary Midgley (1985) draws attention to this characteristic of human language in an investigation of duties that humans might have towards islands. She writes that in philosophy 'duties' are often understood in a narrow legal or political sense, based on social-contract understandings of the political community, while we use the term in a much broader sense in daily life – it makes sense to speak about duties towards children, cats, or one's garden. The narrow social-contract understanding of 'duties' is often portrayed by philosophers as universal, while it is in fact highly specific, and contingent. Narrowing 'duties' to legal or political duties obscures the fact that we use the word in many instances, for many types of duties, including in daily life. Philosophers, Midgley argues, often do not have the right vocabulary for these duties we live by. In order to do justice to these duties in our lived reality and to their meanings in language, philosophers should embrace the many duties behind the word duty. 'We have quite simply got many kinds of duties, including those to animals, to plants, and to the biosphere. But to speak in this way we must free the term once and for all from its restrictive contractual use, or irrelevant doubts will still haunt us' (1985, 41). The view that rights and duties belong together and are only in place between equals, which follows from this understanding, is problematic too, Midgley writes, because duties often matter much more in situations of inequality.

While we should be careful not to depoliticize concepts like rights or duties in relation to nonhuman others, Midgley rightly challenges a specific rationalist human bias in political theory and philosophy. She also shows that it is fruitful to look at everyday language in determining how humans should behave towards others, instead of simply arguing for including them in human systems on the basis of their capacities, which is one specific language-game that is being confused with a universal truth.

Developing a multidimensional model of multispecies conversations requires a sensitivity to different language-games, and investigating their edges, and the points where they might connect and disconnect (Tsing 2022). In the discussion above, I drew on insights from many different language-games such as indigenous philosophy, political animal philosophy, the literary essay, plant science, art, and ecofeminism, in order to point to the different entanglements of humans, other animals and plants, and the concepts we have to describe these. In these language-games, some of which are part of contrasting knowledge paradigms, such as western and indigenous plant theory (Kimmerer 2013), words do not have the exact same meaning. Therefore, all of these language-games offer a specific lens through which we can view the world and relations with others, who may take part in them, or influence them indirectly. Thinking through political concepts in relation to other animals and plants asks for being critical of anthropocentrism and remaining attentive to the agencies of the other beings involved, but also for attending to the multiple language-games hidden behind concepts, that change over time. The meaning of 'dialogue' changes in the multispecies context, and this is not a loss. In rethinking concepts with more-than-human actors, we can draw on the meanings they already have, and explore their potential for guiding new relations.

As Greta Gaard (2016) points out, forming new relations with more-than-humans asks for taking the perspective of others, and learning to listen. At the same time, our world is and always has been multispecies, and there is a lot that we do know about the beings with whom we share the world. Zwamborn ends *The Seaweed Collector's Handbook* (2020, 134) with a sentence from a poem by Monika Rinck: 'At first I saw everything from below, and then I was algae.' We already know more about the worlds of others than we think. Not only about the worlds of animals and plants, but also that of the sea itself.

The work of Astrid Neimanis (2017) provides a radical starting point for better understanding the sea, namely as the watery bodies that we all are. She proposes to take seriously the fact that, as humans, we exist mostly of water and are connected to other watery bodies in ongoing processes of transformation. Understanding embodiment as watery, according to her, challenges 'dominant Western and humanist understandings of embodiment, where bodies are figured as discrete and coherent individual subjects, and as fundamentally autonomous' (2017, 2) and opens up the way to a more fluid and porous understanding of self and relations. Neimanis writes that the 'flow and flush' (2017, 2) of water constitutes who we are – we are constantly engaged in processes of intake (like drinking) and exchange (peeing), that connect us to other water bodies. Viewing humans in this way challenges individualism but also anthropocentrism, and it offers a more-than-human interpretation of the human through water. We are similar to the sea, Neimanis shows. We can learn from it through watching its cycles of ebb and flow, and its relation to time, and its generosity in providing space for others to live. And similar to those of the sea, our borders are porous too. We are always connected to other bodies and beings that influence what and who we are. Recognizing and fostering our porosity can help us live better with more-than-human beings, as the more-than-human beings that we are.

CONVERSATIONS WITH ART

I suddenly knew that I had to write a novel about the sea flooding the Netherlands. All of my novels begin with a vision, which is not a full outline of the story but rather an abstract awareness of what the book is about and its atmosphere. This vision usually includes the beginning, and often also the end. When I began writing this book, which was published under the title *Zee Nu*,[1] I knew the sea would progress at the speed of one kilometer a day. This gave the book a nice pace, my working days too – I wrote one day a day until the story gave me a storm, and things changed. Novels are always a matter of question and responses between me and the story (including its language). I follow more than I steer. But this book was the easiest so far, I only had to sit down and let the story write itself.

City pigeons like art in public spaces, and statues specifically. They use them to rest on and, when they are combined with a fountain, also for washing and cooling off on hot days. For humans, statues have one meaning (or several, as the recent protests to colonial statues showed), for birds they have another, and so statues connect different understandings of the world.

In this book I am not just in conversation with nonhumans, but also with philosophers and other humans, some of them alive and some of them dead. In my life I have similar conversations with artists and writers, and their work, through reading and rereading books, seeing and reseeing works of art, and sometimes responding to these works in my own work. The conversations that I have in my work with living humans are not so different from those with the dead, because their work will answer in new ways, every time I return to it.

1 The English translation of this novel (*Sea Now*) will be published by Peirene Press (UK) and Two Lines Press (US) in 2025.

6. In Dialogue with Art

Abstract: Here I discuss the conversation I have with my work, using novels as an example. This ongoing dialogue forms my days and character, though not in the same way as living with animal companions does. Drawing on insights from new materialism, I argue that the porousness that is required to write shows us something about human agency more generally and matters democratically. I also discuss art as a 'public thing', following Honig's views of public things, to better understand its democratic power. This power also plays a role in more-than-human relations. To explore the role of art in improving multispecies relations, I briefly discuss my work with a multispecies art collective.

Keywords: art, literature, thing power, art agency, human agency, the piano, multispecies art, multispecies collective

The relationship I have with my work is one of the most important relationships of my life. Similar to the conversation I have with Olli, the conversation with my work shapes my days and my outlook on life, and the kind of person I am. As a dialogue, it is located somewhere between the dialogue I have with myself – which I will discuss in more detail in the next chapter – and the dialogues I have with other animals, including humans. Through the work I am also in dialogue with other humans, but that is not my primary concern in this chapter.

In what follows I first describe the dialogue I have with my work, in which I focus primarily on writing novels; similar dialogues take place when I write music, songs or poetry, or draw. I then analyze the porousness that characterizes my agency in the relation with my work, and what this means for human agency more generally. Art not only shapes the artist, it also enters into conversation with larger society, as I discuss in relation to Jane Campion's film *The Piano*. In the second part of this chapter, I focus on the agency of art in human collective conversations, in which draw on Bonnie Honig's ideas about public things (2017). I also explore the role of art

Meijer, Eva. *Multispecies Dialogues: Doing Philosophy with Animals, Children, the Sea and Others.* Taylor & Francis Group, 2025
DOI: 10.5117/9789048564415_CH06

in multispecies collective conversations through a brief discussion of my work with the Multispecies Collective. Finally, I turn to what art teaches us about the different shapes that conversations can have.

The dialogue with my work[1]

Every morning, Olli jumped onto the couch and lay down next to me, after our first walk of the day. Pika used to do the same; he took over the habit after she died. In this first hour of the day, I drink coffee and write. When I work on a novel, I generally begin with writing some pages for that because the day is fresh, in the morning hours, and my view is clear. I can write at any time of day, with any kind of distraction, but this is what I prefer. Later in the day I might work on academic texts, which are less demanding, answer e-mails, and do other kinds of work.

Before I studied philosophy, I went to art school, and before that I studied music. Before that I wrote poems and songs, and performed them live. This

1 While I use the word 'work', what I do is not captured by usual definitions of work. Raymond Geuss (2021) draws attention to the fact that the word 'work' refers to many different types of activities. The general meaning is based on the type of industrial labor that became prominent after WWII, he argues, and is interconnected with a general framework for life, at least in the West. He then distinguishes between three features of work: it is a strenuous human process, it is a necessity of life, and it produces something. He adds that work is a distinct activity (from 'life') and that one does it not for fun. Finally, it is monetarized: people get paid for work. Geuss makes it clear that he does not aim to provide an official definition, but rather wants to capture the general meaning of the concept of work. From Geuss' analysis it is easy to see why people often do not think of art as work. Creating art does not need to be strenuous, although it may involve a difficult working process and always asks for serious concentration, many do not see it as a necessity – art has a function, but a different one than the type of industrial labor Guess describes, and while there can be a product as the outcome (a painting, book, performance, song) this is not necessarily the case. Art is not separated from life but rather intertwined with it and in some senses more akin to play than to industrial labor. Like philosophy, it is a type of practice that reflects on life and its meanings. Work in the sense that Geuss describes will shape the human doing the work too (it may change their bodies, as in physical work, their minds if they work at universities, their daily habits, outlook on life, ethos, and so on). But in the case of art there is a very intimate connection between one's self and the work – artists speak through what they make, and the agency of the work often takes center stage in the life of an artist. At the same time, making something demands reflection and taking a step back from oneself – not letting the ego intrude into the work. While artists may speak through their work, the work only uses the artist to come into being. So, in a sense I do not work, and I do not see writing or making art as work (answering e-mails can be work in the other sense, or doing interviews). At the same time, the activity of working on my work is my job in life, so words like work and job have a meaning for me in this context.

was what kept me going as a teenager and for quite some time after that too. Now I mostly write novels, sometimes poetry, and in recent years I began making visual art again more seriously, also together with nonhumans.[2] Throughout my life, my work has been my companion and safe space. The work requires that I give it the best of myself, which is demanding, but it also gives me a clear sense of direction, and gives my life meaning.

The process of working formed me because I am always in dialogue with something else that has more value than I do. The work itself shapes me through the dialogues I have with concrete projects. This is a matter of feeling as much as thinking, as when the novels usually begin with a vision that comes out of nowhere. When I start writing I usually roughly know what happens, and I know some of the characters. The story, or a character, then asks for something, which leads to something else. I sometimes respond to the text that unfolds by tidying it up, or changing the course a bit, and sometimes I can decide upon what happens or is being said, but my job is mostly to follow what the story asks, without knowing where we will end up. When there is a first version, I can reflect on what I have written and revise it, alternating between the position of editor or reader and that of the writer or representative of the book. The question is never if the book is good, it is not my job to judge that and books only gain their full meaning over time, but only if it is what it is supposed to be. While it may seem as if I make it all up, or intentionally steer the work, my experience is that I have relatively little power over it. This is a form of unselfing, in the words of Murdoch (1971), even when I write autobiographical work. Unselfing is necessary because the ego, or my opinion about the work, would distort it, and asks for practicing an attitude of receptivity. While my agency obviously affects the work – it otherwise would not exist –, the work affects me in a similar measure. I not only write a novel, the novel writes something in me too and I come out of it a different person. A book always carries a lesson for the writer, which is generally personal and different from what it tells the rest of the world, and also private.[3]

In this dialogue the outer world exercises agency too. The relation with my work is always also a relation with the world: when I am working the world discloses itself through me and to me by shaping the work,

2 https://themultispeciescollective.cargo.site/. Accessed November 9, 2024.

3 When I write, the text has a hold over me, specifically the novels or other story-based work. There is a constant pressure to continue, and the work not only takes up space in the hours when I write, but it encompasses everything. Characters sometimes linger around after the book is done.

and when a book, song, or drawing is finished it goes into the world and speaks for itself. At that point its power over me disappears, and I am not interested in it anymore. Some writers enjoy seeing the book they wrote in print, but for me the service to the book is done. While I wish the book well and as a bystander may appreciate it when others connect to it, it has nothing to do with me anymore and its full meaning only comes into being in relation to different readers – books are unfinished until they are read and read again. (We live in a culture in which writers are constantly asked to explain their work, but at that point my opinion of the work is not categorically different from someone else's opinion about it.) The triad of world, work, and self ends when the work is finished, and I begin a new project.

Porosity and the 'and': Agency in motion

Working like I do follows from, and requires, porosity. I let the world flow in and out, and I allow the work to take up space inside me, while I also bring it back into the world by working. This process is something I have learned to trust. I never decided to work in this way but have become better at it throughout the years, and 'better' here means becoming less present in the work.

In *influx&efflux* (2020), Jane Bennett describes a similar ebbing and flowing of the world into the human, in conversation with the work of Walt Whitman. Bennett's concern is how to understand human agency in a world full of thing power. She conceptualizes a form of non-sovereign agency, or in her words: develops a 'model of I' (2020, xi), that does justice to nonhuman influences on human forms of being. In describing Whitman's ideas about agency, Bennett draws attention to the fact that we are not only porous, but that our porousness is dynamic. Part of being an I that is formed by the 'influx and efflux', which Bennett describes as the in and out, the comings and goings, that are encounters, is that the I is always in motion. Bennett describes this I (via McKim Marriott) as 'dividual' (2020, xii) – a self is not an individual single person as a bounded unit, but constantly influenced by heterogenous material forces while giving something of itself back to the world. At the same time, there is an I – metamorphosis, change, or becoming, requires that something remains the same and that something changes. This is true for the influx and efflux as well, in which 'self' and 'other' are porous yet apart.

Whitman's writings about the sea illustrate this process. In the poem 'As I ebbed with the ocean of life', he writes about an I that floats like Long Island (Paumanok) upon the Atlantic (Bennett 2020, 86–87). The sea-self he describes in 'Song of Myself' similarly follows the ebbing and flowing of the water, as well as the waves and the storms (Bennett 2020, 86–87). The human self, which Bennett describes as an I that is 'of one phase and of all phases', (2020, 87), resembles sea beings, which exist in every wave, and all waves. We are similar to nonhuman beings, and if we pay attention, we can learn to align our movements with theirs. In this influx and efflux, Bennett emphasizes the importance of the 'and', and writes that it 'marks the *hover-time* of transformation, during which the otherwise that entered makes a differenee and is made different' (2020, x). Between the in and out, there is an encounter, a moment of change.

Writing a novel and making art more generally is, in my experience, testimony to the 'and', because the work comes into being in the moment between when the world comes in and when it leaves the artist (an individual or collective). The moment of origin or transformation cannot be forced or grasped, but through working I continue to make space for it. Similar to relations with and about the sea in the previous chapter, in this process different types of agencies connect. Humans are powerful in some ways – they can write a novel – but vulnerable in others – this novel will probably survive the human, and the language surely will. As Bennett shows, this assemblage of agencies is continually in motion. In the process of writing a book, my relation with the book changes, and throughout my life my relation to writing changes.

This dynamic characteristic of agency is relevant for my investigation into the edges of conversations. In fact, the process of 'influx and efflux' that Bennett describes works like a conversation: of the world entering and leaving the self, in a continual movement of question and response. This is also why it is so important to see conversations with nonhumans as processes. We need time to get to know one another, and see one another, but we are also always connected dynamically. The movement inherent in the figure of the conversation (of back and forth, question and response, the fact that the 'and' takes time) is connected to transformation and change.

The conversation between the artist or writer and their work may seem highly specific and rare. But art also speaks to who sees, reads, or listens to it; to individuals and communities. Art co-determines how humans see themselves, and has the power to challenge that. Before I discuss the agency of art in the world in more detail and turn to its democratic function, I want to linger on this strange power and what it tells us about reaching and moving others, through discussing the role of the piano, and of music, in Jane Campion's film *The Piano* (1993).

The piano as a person and the force of art

Ada, the human protagonist of *The Piano*, does not speak. Instead, she plays the piano, and this instrument can be seen as the nonhuman protagonist of the film. The story begins when the Scottish Ada is sold by her father to marry a man she does not know in New Zealand, in the mid-1800s. She brings her nine-year-old daughter Flora, and her piano. Her husband Alasdair soon sells the instrument to the neighbor, George. Ada is upset, but George asks her for piano lessons. He tells her she can have the piano back, key for key, if he can do 'things he likes' when she plays. This proposal leads to an intimate relationship, because in contrast to her husband, George listens to Ada, and hears her. After a while, George wants her to care about him, and not be with him because of the piano, so he gives her back the instrument. When Ada visits him because she misses him, her husband finds out, locks her in their house, cuts of her finger, and then understands he cannot possess her and should let her go. In the last scenes of the film, Ada follows George to his new home, on another island, with Flora. They bring the piano too, but when they are on open sea, she tells him to throw it overboard, because it is too heavy. When the rowers finally oblige, she puts her foot in a knot in the rope, so that it drags her with it to the bottom of the sea. Something in her wants to live, and she kicks herself free. The film ends with Ada learning to speak and giving piano lessons, living happily with the man she loves.

If one only focuses on the humans, the film can be read as sketching different kinds of ambiguous power relations. For example, Ada is subjected to patriarchal oppression but has her own kind of power, and the men in the film suffer too because of the oppressive social structure. Flora is a child, and not very powerful, but she does represent her mother, speaks for and about her, and betrays her too. The Māori are figuring out how to deal with the white men stealing their land, but have their own force and are more comfortable in their lives, landscape and nature than the white people in the film.

Power is not only connected to language, but also to silence. This is most clear for Ada, who chooses not to speak anymore when she is six years old. Her father calls her stubbornness a 'dark gift' and says that the day when she chooses to stop breathing will be the last day of her life, referring to her strong will. Her chosen muteness represents the larger context of her life, in which Ada does not have an actual voice – her father and then her husband decide for her. But not speaking is a form of resistance too, and through playing the piano she does speak, but in a language of her own choice. When George recognizes this, she can finally choose to live a human life, with other humans, which does include the piano, but not the strangeness of it, the otherness. This strangeness does remain inside her, which becomes clear in the last scene when, in a voice over, she speaks about the piano at the bottom of the sea.

Only focusing on human agency, however, does not do justice to the story, or to the nonhuman forces in the film. Ada's dependence on the piano shows that not only humans exercise power, and throughout the film many of the characters are strongly affected by nonhuman forces: the music that Ada plays provokes strong emotions in her husband, the people of the town, and George. This music is wild and beautiful, but in a strange way – 'she does not play like us,' other women say, which is another way of saying: she is not like us. Ada herself needs to play, her relation with the piano is symbiotic, and when she is separated from her piano, she carves out piano keys in wood to play on. There is something more at stake than what is normally understood as music, playing the piano is a form of being, or freedom.

Other nonhuman forces are also portrayed in detail. The sea, the trees, Ada's clothes, human bodies, and stories that characters tell one another (or stage as a play) shape the deeds of humans in the narrative, but also visually. Fingers are compared to trees, the piano is made of dead wood, the sea expresses the force of the music – visual rhyme is central to the film, and there is an emphasis on tactile and sonic expression. Sometimes the camera follows the perspective of Flora or her dog, making the viewer take the perspective of other animals. Close ups also add to this.

While the emphasis on thing power – the power of nature and art – can be read as paying homage to the Māori in the film, their agency is underexplored (even though their refusal or cooperation is sometimes essential for the story to unfold), and the same is true for the nonhuman animals in the film. The trees are taken more seriously. But the film does show that there are many ways of speaking, and many ways of being silent. Characters speak through words, music, body language like withdrawal, eye contact, gestures, touch, sounds, and in other ways.

One of the most important alternatives for human language in communication is music. In this film both the piano and the music are full characters. The piano is sometimes used to speak for Ada – alone on the beach it portrays her loneliness, close ups of keys when she plays for George suggest touching him. But it is also a piano, an instrument that is respected and feared. For Ada, the piano is her partner, and while their relation is different than the relations she has with George or Flora, it matters just as much.[4] The music by Michael Nyman is foregrounded in the narrative.

4 Kevin McNally (2019), who writes about the relationship between gamelan instruments and players, calls instruments 'duet partners'. In Java, the instruments in a gamelan orchestra are

Other piano films, like Haneke's *The Piano Teacher*, or Polanski's *The Pianist*, use music to complement or enhance the human-focused narrative, but the main focus is on the humans, the music itself is an ornament or illustration. In *The Piano*, the music moves the story forward and speaks in its own way to the viewer/listener. It takes up space prominently, both in the story and in relation to the visuals, that sometimes seem to work to accompany the music and not the other way around. The human characters need to deal with the power of the music, and the freedom it proposes. This power shapes not only their individual modes of being, it also affects the society – the women who comment on the strangeness of Ada's playing exemplify this, in their outspoken disdain of the music, which contains a desire to be like her. Similar to nature, the music has a force of its own, that transcends human modes of being in a way that is not religious but does connect to something other.

Public things and the democratic power of art

When we listen to music or view a painting or read a novel, the work enters into a dialogue with our life story, by affecting us as individuals. But art also affects societies. The community of which Ada is part is changed in several ways by her music, not just aesthetically but also politically, because her playing defies gender norms and expresses a kind of freedom that other humans envy. In what follows I explore this strange power of art, through the role that works of art plays in social and political dialogues. For this I first turn to Bonnie Honig's (2017) view of 'public things'.

Public things are objects used in common, such as schools, rail roads, sewage systems, public telephones, which play a role in holding a democratic community together (Honig 2017, 4-5). Honig writes that they provide an infrastructure for interaction which gives citizens something to speak about, and provides them with the 'holding environment' necessary for democratic citizenship (2017, 5). To elaborate on the role that public things

seen as persons, and the gongs are the soul of the ensemble. McNally developed a performance for a human performer and two Javanese gamelan gongs, which investigates the relation between human agency and instrument agency. He argues we should 'tune into' the other, instead of following schemes of economic rationality, and that this will allow us to better perceive the agency of nonhumans, and blur the subject-object divide. McNally proposes to take sound instead of vision as the basis for knowing and being. Because sound cannot be separated from its environment, this can teach us ecological sensitivity too. Similar to composer Pauline Oliveros, whose work I will discuss in more detail in the last chapter, McNally argues that developing (ecological sensitivity) through attending to sound is not simply an exercise of the mind, it asks us to actually listen.

play in democratic life, Honig draws on Hannah Arendt's threefold distinction between types of human activities: labor, work, and action. Labor, for Arendt, refers to activities that are necessary for self-preservation and the continuation of the species, such as harvesting food or housekeeping. These activities need to be endlessly repeated for human survival. In contrast, work leaves behind a durable object, and as an activity is demarcated in time, it has a beginning and an end. Use objects like a table or a spoon are products of work, and artworks are too. The last category of Arendt's human activities is action, which refers to acts in which humans show themselves to others in the public sphere. In action, one reveals who she is to others, and through action humans encounter each other as unique and distinct beings. Action is unpredictable, and for this reason it makes possible the new.

Honig draws attention to the fact that work provides both labor and action with the necessary stability. Because of the emphasis on action in Arendt scholarship, Honig writes, scholars have overlooked the importance of work (2017, 41). But for action we need a space in which we can act, and in the current neoliberal political climate, we find out what happens if the public sphere disintegrates through the erosion of public things (Honig 2017, Lecture 1; Brown 2015). To counter this, we need more attention for work.

Art as a public thing

In her description of public things Honig does not explicitly discuss the status of artworks as public things. She mentions art only once in relation to Arendt (2017, 39),[5] even though she does use works of art such as Lars Von Trier's film *Melancholia* to develop her views. Viewing art as a (collection of) public thing(s) can, however, bring out how art affects social and political relations and creates a common world, as well as how it shapes collectives, and democratic dialogues.[6]

In her discussion of the role and power of public things, Honig makes it clear that she does not accept a vital materialist reading of objects, in which the things themselves have agency and the human is seen as part

5 I should note that in Arendt's view, art can be both thing and action. While art institutions such as museums and concert halls, as well as public art like statues or community art projects, may function in ways similar to libraries or schools – they provide a space to gather and to speak about – there is also art that resembles action, such as theater and performance.

6 With regard to the role of public things in neoliberal societies, there is furthermore a uselessness to art that explicitly challenges neoliberal values, making the creation of art an important form of critique. While some art has economic value, the value of most art – including whole genres, like performance art or poetry – is difficult to measure in this neoliberal-capitalist understanding of value. In other words, art is useless, not because it has no use, but in that it is meant to be useless in neoliberal understandings of use, because it expresses other forms of value.

of a larger constellation of forces (2017, 28). Rather, she focuses on human-thing *relations*. She reads the work of Arendt in conversation with that of Donald Winnicott to argue that things 'enchant our world, inhabit us, and press us into object-mediated relations with each other and with a world of things' (2017, 7). While the resistance and power of things co-shapes human subjecthood and our lifeworld, the human 'remains the focus' for Honig (2017, 28), to which she adds that her concern is political rather than ontological.

This analysis of how objects shape relations can help clarify the importance of art for societies. Art plays an important role in social and democratic life by responding to power relations in society, twisting existing narratives, articulating other ways of being than those that follow the norm, abstracting from reality, and challenging and fostering our collective imagination. As public things, works of art give us something in common, to speak about, and in the case of museums or theaters also a space to speak in. Art can ask humans to view their life and society differently, and can shape the meaning that humans give to their collective life.

However, as a public thing an artwork works differently from a railroad or a hospital, because works of art do not just facilitate discussion, but often also provide a critique of society, and because they help us imagine social and political alternatives to the status quo. In other words, art does not just provide solidity for the common world we share with others: it also has the power to criticize the understanding that individuals and communities have of themselves. Art can touch us and move us, and show us that the lines we draw – around 'self', 'human', 'community' and other categories are, in fact, porous and open for debate.

Public things through a multispecies lens

While Honig convincingly shows that we need public things for a democratic holding environment, and while her ideas about public things can help us understand the role that art plays in societies, for her 'society' means human society. The understanding that we always live in a more-than-human world is lacking from her analysis. Honig does not discuss nonhuman animals even though many of them are part of human communities too, as workers, companions or neighbors, and she does not elaborate on the importance of multispecies holding environments that are under threat or destroyed by neoliberal capitalist forces for other than human beings.

This is problematic for several reasons. Societies are already more-than-human, as I showed in earlier chapters. Similar to how this works in the human case, for creating solid multispecies relations we also need spaces in which we can speak with one another (Donaldson 2020, Meijer forthcoming b), and we need things about which we can speak to one another. Many of the public things that Honig mentions can or do also play a role in multispecies relations. Infrastructure, healthcare and education all matter for nonhuman animals too, and have a role to play in multispecies communities (Donaldson and Kymlicka 2011). If we keep centering the human and human relations in discussing the value of public things, as Honig proposes (2017, 28), we risk repeating anthropocentrism. This favors humans at the cost of other members of other species, which leads to exploitation of nonhumans and makes it impossible to solve the cluster of ecological crises we are facing. It also marginalizes humans by favoring a specific type of human (Ko and Ko 2017; Taylor 2017, see footnote 48). Setting up a binary between 'thing' and 'human' furthermore obscures the fact that there are things and humans in the plural, and many other agents too (like nonhuman animals). There are myriad relations between different entities, in which different kinds of agency play a part. Similar problems surface in Honig's idea of 'world', as one (human) lifeworld (see for example 2017, 38–40, and 2017, 1–3). This understanding of 'world' obscures that there are many worlds, shaped differently in different human and animal cultures, with their own epistemologies and cosmologies (Escobar 2020), and their own needs for public things.

In working towards more just and sustainable multispecies communities, we need to take seriously multispecies, and nonhuman,[7] public things. It is not up to me to provide a full list of public things from behind my computer. As Olli, the mice, and the amphibians showed in earlier chapters, further exploring multispecies relations and working towards ecopolitical change asks for careful engagement with others. The same is true for mapping multispecies public things and their role in the many worlds that make up our world. However, I do want to point to one form that multispecies public things can take, which is multispecies art.

Art and fostering multispecies dialogues[8]

I am part of a multispecies art collective, called the Multispecies Collective,[9] together with my friend and human artist G. C. Heemskerk, and our dog companions Olli and Doris Meijer, and Miemel and Wiske Heemskerk. The collective is an artistic multispecies collaboration but also a research

7 For example, humans now often destroy nonhuman animals' infrastructures, like routes that elephants have taken for generations, which is for them clearly a public thing in Honig's account.

8 This section contains some insights I explore in the essay 'Tierkunst als weltbildende Praxis' (2024).

9 https://themultispeciescollective.cargo.site/. Accessed November 9, 2024.

project in which the humans look for and document nonhuman animal and plant art.[10] Our audience is also more-than-human. In 2020, we curated an exhibition in my garden, called *Verwerelden* (Reworlding).[11] In 2022 we were artists in residence at the Pompgemaal in Den Helder in the Netherlands. For a month, we worked on location in the dunes, looking for animal art and respectfully introducing our work to the nonhuman inhabitants of the area as well as a human audience. We also organized dog dinners,[12] vegan dinners for companion dogs and their humans, developed in collaboration with dogs, and wrote a Multispecies Manifesto for justice in art institutions.[13]

Contemporary artists increasingly recognize and challenge the limits of human agency (and the porosity of humans), and the power of nonhuman agency (Gibson and Sandilands 2021, Page 2020; Moran 2022; Ullrich 2019, 2022). There is also increasing attention for the role of power relations in determining what counts as art, which is interconnected with a critique of individualism and hero worship in the art world (Zarobell 2022). Art collectives such as ruangrupa or Array Collective argue that art is not made by autonomous geniuses, but by beings who are always formed by situated and embodied relations to other beings, and larger social, economic and political structures. Artists depend on others for material, like paint or paper, space and time to work, and for ideas – all art builds on other work. These collectives aim to provide an alternative to capitalist ways of making art.

With the Multispecies Collective we also want to challenge existing norms about who counts as an artist, and what counts as art, but we do so in recognition of the fact that we live in a multispecies world in which different beings create meaning in their own ways. Even though there are many art forms that play a role in nonhuman animal lives and communities, such as performance, dance, land art, readymade, song, sculpture, or architecture, humans hesitate to call these forms of expression art (Gigliotti 2022; Ullrich 2022). For example, animals who build nests are often thought to act on instinct, while human building practices, according to humans, require intelligence. Or animal buildings are seen as simply functional, where human buildings are seen as also aesthetic. But a refusal to view animals as artists is not just connected to doubt about their capacities: humans are also the ones who define what art is, and generally define it as human (I discuss this

10 https://meersoortigcollectief.blogspot.com/. Accessed November 9, 2024.

11 https://verwerelden.blogspot.com/. Accessed November 9, 2024.

12 https://www.mediamatic.net/en/page/388941/dog-dinner. Accessed November 9, 2024.

13 https://themultispeciescollective.cargo.site/multispecies-manifesto-en. Accessed November 9, 2024.

in detail in Meijer 2024). However, if we take a closer look at what is meant with 'art' we find it refers to many different language-games, which do not share one characteristic and which change over time; what we now see as art is very different than what was seen as art a hundred years ago. Many of these language-games are (co-)shaped by nonhumans, and there is no clear cut between all human art on the one side and all animal creative projects on the other. There are many different forms of art and design, in which the agencies of nonhumans may resemble that of humans or totally be different (Dufourcq 2021; Gigliotti 2022; Sandilands and Mortimer 2021). For example, elsewhere (Meijer forthcoming a) I propose to view Olli's digging of dens – a learned skill that requires aesthetic insight – as a type of practice which is located somewhere in between art, habit, and self-expression.

Art expresses and creates meaning in a given life-world, and this is not only true for humans but also for other animals, who for example connect and communicate through dance, song, building nests or burrows, or other interventions in the landscape. Further understanding of how this works asks for attending to the practices of others, and respectfully engaging with them in common practices, if they want to. For the Multispecies Collective, this, for example, means that the working process matters more than the final works of art, and that listening and attending to others is as much part of the process of creation as our self-expression, or human ideas about the world. How projects evolve is as much up to the others as it is to us. Working in this manner changed my aesthetic views. For example, I now really appreciate the shapes of different holes in the ground made by worms or insects, where before they all looked similar.

Multispecies art can facilitate human conversations about nonhumans and art, and can function as a public thing in human society. But the working process we are committed to also involves many conversations with nonhumans, and the art works we encounter and create can sometimes be seen as multispecies public things. Artworks can include the digging of holes, playing back sounds from the environment, making new paths by walking, mutual observation, or the construction of new habits. Moving beyond human frameworks requires the imagination just as it requires critical analysis, and other animals, as well as human artists, can help us make this change.

Because much art that is made with nonhuman animals and plants takes something from them without giving back (think for example about art installations in which more-than-human beings are used), we currently explicitly focus on giving back. G. C. Heemskerk is working on a stringfoot project with pigeons in The Hague, and we plan to make a film about the amphibian migration, to make their vulnerable position visible to other humans.

Art and the many shapes that conversations can have

A focus on art shows that human conversations can include many different forms of expression. Humans can speak to each other in music, images, and other forms of expression that do not rely on human words. Art can also bring human conversations into being: it can bring people together and offer them something to speak about, which can affect the self-understanding of human collectives and communities. This has democratic value. Art can play similar roles in multispecies and nonhuman communities, because humans are not the only animals who create meaning, or who contribute to shared life-worlds with others creatively.

The dialogue I have with my work is an embodied and situated dialogue in which different senses play a role, and in which my agency is connected to the agency of others. In order to speak better with the work and other-than-human beings, unselfing is important. Dialogues with and about art share resemblances to the conversations with and about the sea, with regard to agency, and the use of different senses. And in connection to art too it is important to distinguish between different types of agency (Meijer and Bovenkerk 2021; Knudsen 2023). A dog who intentionally works on their dens each day takes part in a very different language-game than a painting that is being looked at or a novel that comes into being through a dialogue that the writer has with their book.

One important characteristic of deliberation that I mentioned before is its connection to transformation: when we engage in conversation with others in an open and honest way, we risk changing (Young 2000). This is true for conversations with art as well. Artworks can become friends, and they can change their meaning for us over time, as we change and encounter them anew. Taking seriously these conversations therefore not only matters for us as individuals, but also as (more-than-human) communities facing great ecological and political problems.[14]

In any case, I am shaped by my work. But the conversation with my work is not the only inner conversation that I take part in. The conversation with my work, and the other conversations I discussed so far, are all accompanied

14 I do not mean the 'message' of art, which can amplify propaganda as well, but its capacity to let us see who, what and where we are differently.

by a dialogue I have not yet paid attention to: that with myself. This dialogue too raises questions about language and agency, and speaking with oneself is thoroughly interconnected with speaking with others. In the next chapter I therefore turn to speaking with myself.

CONVERSATION WITH DEPRESSION

A winter morning. I stand in the kitchen and make coffee. I think: can I go yet? No, I tell myself, you cannot go yet.

7. Speaking with Myself about Depression

Abstract: In this chapter I discuss the dialogue I have had with myself about depression in the past thirty years. I contrast this dialogue with Hannah Arendt's view of thinking as a dialogue between I and I, in order to develop a more embodied and situated view of thinking. I also investigate how conversations with the self and with others relate in the context of depression, analyzing how (historical) power structures, and the (medical, political, existential) lens that we use to describe and treat it, can affect how we think about depression both collectively and individually. As a conclusion to the chapter, I discuss how an embodied view of thinking can change how we do philosophy in a multispecies world.

Keywords: philosophy of depression, inner dialogue, mad studies, embodied thinking, democratic thinking, Hannah Arendt

Thinking is always thinking with others but it is also always thinking with oneself. In *Life of the Mind*, Hannah Arendt describes thinking as a dialogue (2021, 241). She calls this dialogue 'traveling through words' (2021, 241), even though it is performed without sound, and moves faster than speaking. Because thinking follows this process of questioning and answering, it can become critical and dialectical. When we speak with others, Arendt writes, the duality necessary for thinking disappears and we become one towards others. The dialogue that takes place in thought, however, does resemble the plurality that exists between different people: the person we are is also other (2021, 240) which is shown by our thought-dialogue, in which 'I' becomes 'I and I' (2021, 236–52). In the thought-dialogue, the aim is not truth, but being consistent with oneself. From Socrates Arendt borrows the idea that this consistency is a form of friendship with oneself (2021, 246).

In this chapter I use Arendt's ideas about thinking as a dialogue that we have with ourselves as a starting point for conceptualizing a dialogue I have had

---

Meijer, Eva. *Multispecies Dialogues: Doing Philosophy with Animals, Children, the Sea and Others.* Taylor & Francis Group, 2025
DOI: 10.5117/9789048564415_CH07

with myself for a long time, about depression. This ongoing dialogue has shaped my view on life, and formed my days. In contrast to Arendt, who describes the dialogue between I and I as something that takes place in the mind, in the case of depression speaking with oneself is embodied too and does not necessarily dissolve when we engage with others. The dialogue I have with myself about depression is a good example of an embodied and situated dialogue, because it does not travel solely through words: changes in my mood influence my thinking in wordless ways and vice versa; meditation, running, and yoga can sometimes change my thoughts more strongly than thinking with words.

The dialogue I have with myself about depression is connected to the world, and engagement with others, in several ways. This is not just my individual experience: discussing one's inner dialogue is part of social practices that concern depression, in dialogues with friends, but especially in therapy settings, where patients are asked to explicate and share the dialogues they have with themselves. For example, in cognitive therapy, patients are taught to steer their thoughts in a certain direction through learning to differentiate between different types of conversations and thoughts.

While depression may seem to be a clear example of a distorted relation with oneself, which influences the kind of conversation we have with ourselves, analyzing how we speak with ourselves in the case of depression also offers more general insights into thinking. To explore this, I will return to porousness, relations with others, and unselfing in conversations more generally. Extending the dialogues we have with ourselves to include the movements of the body, and conversations with others, is also relevant for rethinking thinking in multispecies relations.

Depression as a phenomenon is philosophically interesting, because it connects to existential questions (Camus 2013), as well as to political and social structures (Cvetkovich 2012; Fisher 2014). However, in this chapter I do not want to focus on analyzing depression. Earlier, I have written about the phenomenon of depression, also using my own experiences as a starting point (Meijer 2021b). I will not repeat these ideas here, but instead concentrate on the dialogue I have with myself about it. There will, however, be some overlap between earlier work and this chapter because the same protagonist is used to clarify what is at stake.

Dialogues about depression

Some memories of my early childhood have the color of depression, or rather the lack of color that I associate with depression. But the first time it overwhelmed me I was fourteen years old. It felt as if gravity was too strong and dragged me to the earth, and made moving difficult. My mood was low, and my thoughts governed by self-hatred. The experience felt very true: life is without meaning, I am without value, in the greater whole, others may not see this, but this is how life is. I read Sartre's *Nausea*, and understood the feeling of nausea; I read Descartes' *Meditations* and felt that yes, we truly are all alone.

In the years that followed, depression overshadowed my life. I spoke with therapists, used anti-depressants, and was hospitalized for anorexia caused by depression. I had good friends, and my work helped me to get through the days – writing poems and songs, singing on stage. The weight of depression gradually lifted, and when I was in my early twenties, life became more bearable. I went to art school, where I learned to work, and played a lot of music. The conversation with the other side of life, which was a continuous presence during my adolescence, became one conversation amongst others. There was no moment of healing, no sudden lifting of the weight, but a gradual change, and throughout my life this change has been something I have come to rely on. In the years that followed, depression has moved around in my life: it has been fully present, near, in the background, and fully gone. Through experience, I have found out that my best response is not to argue with depression, but to keep moving physically, and to focus on what is not me – my work, activism, or nothingness (for example in meditation). This is not the same as running away from my feelings, or ignoring the state that I am in; it is a way of living through episodes.

In the past thirty years, the conversation I had with myself about depression has taken different forms. I have had psychotherapy and cognitive therapy, where the kind of dialogue that Arendt describes is encouraged. Going back and forth mentally between different views of yourself and your life is used to teach people to bracket unhelpful thoughts. Explicating this dialogue and questioning the truth value of ideas I had about myself did help me when I was younger. But in my day-to-day life, embodied responses are more useful. Working moves my focus from how I feel-think to something outside of me that has value, regardless of whether I have value or not, and the dialogue I have with my work is far more interesting than the dialogue I have with myself about depression. There have been times in which my work kept me going. Not just because it keeps my mind busy, but also because through drawing or playing music I can refigure where I am. Walking and running also helps (Oswald et al. 2020).

In addition to these practical interventions, I learned that a sensitivity to melancholy is a part of me, and while I cannot say that I have gotten used to it, or reconciled myself with it, I have gotten used to life and myself over the years. In the process of getting used to life and finding a way to live through days, my animal companions have played an important role, both in a practical sense, for example when Pika taught me to take long walks, as well as existentially because they show me what matters, like the mice do in their caring practices, or Olli when he sat next to me in the morning. I also learned not to expect much from my own life, which is not sad but liberating (even though I might sometimes feel sad about it).

What the dialogue about depression I have with myself tells us about the types of dialogues we can have with ourselves more generally

Oversimplifying, Arendt's dialogue between the I and I has the shape of a conversation between two humans who use words. If an actual other enters into the picture, the I and I melt into one self, and the I has a similar conversation with another human. Some of the conversations I have with myself follow this model, including some of the conversations I have with myself about depression. Sometimes I use rational arguments to end a line of thought, or to change my view of myself or a certain situation. Sometimes I can also lift up my thoughts mentally and move them in another direction. But just as often, I respond with my body. Going for a walk might be a rational choice – I feel bad, I should go for a walk because I know that afterwards I will feel better – but the actual answer to depression is in the process of walking, the movement of the body, the view of the landscape around me, touching the earth with my feet. Furthermore, unless I feel really bad, I do not make a rational choice to go for a walk in order to feel better: it is simply a habit that I have, and my body (or Doris) will get up around a certain time to do it. Rather than resembling a conversation in words with humans, the dialogue I have with myself about depression is an expression of *sentipensar*. Sentipensar is a concept that anthropologist Arturo Escobar describes as thinking-feeling (2020, 2), two activities that in different cosmologies of indigenous peoples of the southern Americas cannot be separated, and continually inform each other. Furthermore, my dialogue does not just involve thought-feelings, but also actually moving my body. As Merleau-Ponty writes (1962, 143; Weiss 2006), aligning oneself with the world is done by the body too, for example through creating habits (1962, 143).

The dialogues we have with ourselves are always located in a specific body, constituted by the dynamic interrelation between body and mind, and located in time too – they have a history, and will change course over time, as when building a habit, but also through new experiences, reflection and so on.[1] For me, the dialogue about depression changed very much throughout the years. I have learned to not take a certain type of thoughts seriously, even though it is harder to not take the feelings that are part of them seriously, and do not even discuss them with myself anymore because I find them too far away from where I am now. Silence is an important part of the dialogues too. When I focus on my work – writing, drawing, playing the piano – I respond to a thought-feeling by moving the focus away from myself. Instead of using arguments, as I would in the type of conversation Arendt describes, I turn away from the depression and leave it to itself. Similarly, meditating also postpones the conversation and moves the focus from the dialogue to silence.[2] This is a gesture that creates space. The dialogue I have

1 Time also is a factor in the dialogues we have with ourselves more generally, because dialogues take place in time and change over time, and because time gives depth to the dialogues that we have. In *Shibboleth: For Paul Celan* (2005), Derrida points to the resemblance between time and language. Using the figure of the date as an example, he draws attention to the simultaneous recurrence and non-recurrence of a day in the year. Dates connect events through time and space, and separate them. Nothing ever happens again, yet, within the framework of the calendar it does. We encounter events through time, through given dates, in remembrances, in what we do and do not remember, collectively and individually. Within languages we encounter words differently whenever we use them. Like dates in time, and the self we are on these dates, words return and never return. They mirror our loss, and even when they describe it, their own meaning moves away from us. Language does not belong to us, nor does time, nor do we. Yet we encounter ourselves again in language and time, and both language and time hold us. Similarly, we encounter ourselves again on certain dates, as self and stranger – the new encounter through time makes us see who we were, moves us away from old selves, and this enables other dialogues. The fact that we can meet ourselves again already shows that we are not the same as we were – otherwise there would not be an encounter. There is always a loss inherent in this meeting. We are not who we were, we lost that person forever. Even if we are now in a better place, we cannot even stay with ourselves. This is why birthdays can be difficult, and holidays, and the turning of the years. This melancholy is also part of the dialogue about depression, which for me is more or less ongoing; as I mentioned it has been absent for longer periods of times, lingering in the background in other times, and sometimes fully present. The fact that it returns makes the dialogue easier – I am more used to life and myself, and to the comings and goings of depression. But it also makes it sadder, and the feeling of depression more difficult to resist, because of the earlier versions of myself I meet, the losses I encountered in time, and depression's recurrence.

2 Meditation can replace one's inner dialogue with absence, silence, and nothingness. The first step of this process is viewing your inner dialogue as dialogue: zooming out and seeing the dialogue from a distance, which is not dissimilar to how we consider our thoughts in cognitive therapy, or to the bracketing of experience in phenomenology. But where the goal in cognitive therapy or phenomenology is to analyze thoughts or experience critically, in meditation one lets go of, or sits with, the thoughts. Letting go is not one act, but a practice, which one can get better at through training. Becoming better at bracketing the dialogue with oneself helps in the case of depression too, because it creates distance. Here as well, insight comes through practice and not through psychological or critical analysis. Replacing the center of attention

with myself is more similar to the dialogues with nonhumans I described in earlier chapters than Arendt's dialogue in words, because it includes embodied interventions, habits and silence.

The dialogues I had about depression with fellow humans call into question another aspect of Arendt's description of inner and outer dialogues, namely that we become one again in relation to others. In talking therapies, the therapist will often invoke a distinction between the healthy-rational self and the depressed self. The rational self is called upon to act in a certain way, for example, to doubt the truth value of certain statements about oneself (like: I am a bad person) or to bracket them, in order to make the non-depressed voice stronger than the depressed one. Furthermore, the dialogue between the I and I can also be amplified or polarized by the dialogue we have with the outside world. Many different authors have shown (Cvetkovich 2012; Fisher 2014) that depression often follows social, economic and/or political inequalities. Our self-understanding is continually formed by dynamic relations of which we are part, and the dialogue we have with ourselves is too, because we are porous beings. The dialogue with the outside world can affect our inner dialogue in a multitude of ways. Not speaking about depression, for example because of the stigma attached to this condition, can result in the experience of a duality towards oneself. Acting normally while experiencing despair can disrupt someone's relation and conversation with herself further.

Finally, depression can make the dialogue between the I and I impossible. When a depression intensifies, having a dialogue with oneself becomes increasingly difficult. This has a temporal dimension and a spatial-embodied dimension. In my experience, temporally it is as if both I's become stuck in a viscous present. I become detached from my former self and from what formerly gave meaning to life, as well as from the future (Aho 2013; Meijer 2021b). Thinking needs a solid past to lean on, even when that past is constantly reinterpreted in the present, and the promise of a future. When you are stuck in the present you lose the will to speak with yourself,

from something (or someone) to nothing, can alter the dialogue you have with yourself. Not because it changes the knowledge you have of yourself or how you judge yourself, but because you understand your position differently in relation to what is around you, and perhaps learn you do not matter so much. Practicing 'sitting and forgetting', as it is called in Daoism, also helps with a specific problem that arises in the dialogue between I and I, which is that it can become too loud and repetitive. This can happen in depression but also in other circumstances – when we worry or obsess over something. In these cases, knowledge and rational insight can play a role, but there are limits to what thinking can establish, and not thinking might be more beneficial and interesting.

not just cognitively but physically too. Thoughts become repetitive and movement becomes difficult – sometimes actually causing you to slow down. This heaviness, which feels like being filled with sand, or a too strong pull of gravity, affects the body and the mind, makes thoughts heavy too, until there is no conversation possible anymore. This not only happens in depression: other strong experiences, like mourning, also affect our capacity for dialogue with ourselves, and show that in our dialogue with ourselves body and mind are intertwined.

Porousness, agency and rationality

When philosophers describe human agency solely in terms of autonomy and reason, I often wonder if they ever fell in love. Like falling in love, depression also challenges rationalistic and atomistic interpretations of agency. In earlier chapters, I drew attention to the non-sovereignty of our agency in relation to nonhuman and human animals, plants, the sea and art. Human agency is always entangled with the agencies of others, who influence us and whom we influence, and whose attitudes and acts co-determine how or if we can express ourselves. This is clearly visible as well in the case of depression: social structures, chemical processes in the body, a sense of humor, animal companions and many other elements influence how (or if) I can act when I am depressed. Whether somebody will get depressed, and if they will recover or not depends on many factors.

Before, I wrote that the different modes of being, knowing, understanding, and reason of different species, (nonhuman) cultures, and (nonhuman) communities have their own wisdom. Dogs and mice have forms of reason and emotion which are not inferior to human ones but different (and sometimes similar). Within human modes of reason, feeling, or sentipensar, we also find many differences, on the individual level, in cultures, or social groups.

This raises the question if depression also offers a specific form of insight, knowledge or even wisdom (as Kusters (2020) writes about psychosis). Being depressed is not enjoyable or enriching for the individual, but it does provide the person experiencing it with a new perspective on life. While the experience of depression differs between individuals, for me it does bring with it a sharp view of certain existential dimensions of human life and death, concerning our mortality and loneliness, which for example has the advantage of showing how insignificant certain daily worries are, or the opinion that others may have about me. Being depressed, however, also distorts the dialogues that one can have with oneself, or erases the capacity for dialogue, and makes it difficult or even impossible to live. So, while depression can give insight into certain dimensions of our being, being depressed is in itself not a good experience, similar to being ill.

As a lens on life, depression is not the opposite of reason, nor is it located outside of reason. The borders between depression, or madness more generally, and reason are porous in individuals: madness can be temporal, people can combine mad and reasonable traits, and some aspects of depression are rational (like the understanding of one's insignificance that often accompanies depression, or insight into one's existential loneliness). The borders between depression or madness and reason are also porous in societies: certain behaviors were once seen as mad and are now considered to be normal responses to unhealthy conditions (Foucault 2013). Throughout history, the concepts 'reason' and 'madness' have been defined in different ways in relation to each other, under the influence of power structures, cultural norms, and changing views of the human (Foucault 2013).

How 'reason' and 'madness' were constructed philosophically is the topic of a debate between Derrida and Foucault about Foucault's book *The History of Madness* (2013; De Ville 2010). In this book, Foucault argues that what we currently call 'madness' has a beginning point in history, namely the classical age (1650–1900). But Derrida argues convincingly that the history of philosophy can be read as a history of conceptualizing reason, or delineating *logos* (Derrida 1978). In this history, reason and non-reason are co-constituted differently in different areas and cultures, and modernity has its own interpretation, but this does not constitute a break that is categorically different from other breaks. A second objection that Derrida raises in response to Foucault's project concerns his method and aims. Foucault uses what he calls an 'archaeological' method, meaning that he focuses on objects and other remnants of the history of madness to let the voices of those who are silent in history – the mad – speak. Derrida writes that there is an inherent tension in this project, or even a paradox, because it uses the language of philosophy, a language-game in which reason is very important, to capture what cannot be said in the language of reason or modern rationality (Derrida 1978). In other words, he questions if a history of silence, of what cannot be said in a certain discourse, can be written in the (philosophical) language of order and objectivity. Third, Derrida problematizes the hegemonic nature of Foucault's enterprise by drawing attention to the fact that the power of reason was never complete. The silence, madness, or 'other side' of language has always haunted reason, and in fact, has always been a part of it (1978, 54).[3]

In this discussion Derrida not simply draws attention to the contingent character of the meaning of concepts, which of course also matters for thinking about speaking with yourself in the context of madness, as the discussion of the political and social dimensions of depression made clear. He also wants to show that identity and non-identity always come together, in any movement of defining. There is not one moment of constitution of madness and *logos*, or reason, Derrida argues. Like silence, madness has always been part of philosophy's discussion of rationality (1978), both as a historical given, but, more importantly, because concepts always carry with them what they are not. How precisely this takes shape in a given era and place is not given, but Derrida emphasizes the ongoing movement which constitutes the inside and outside.

In other words, Derrida shows that concepts are porous. Throughout history, the relation between 'madness' and 'depression' to concepts like 'normal', 'healthy' or 'rational' changed. There are no universal binary oppositions between concepts, they shift, and the lines between concepts are continually redrawn. There is also porousness in the process of defining itself, apart from any specific historical era: concepts are never established neatly, their meaning always varies between language-games and even individuals, and concepts rely on other concepts for meaning anything at all.

Following Derrida's analysis, we see that what is mad and what is not in the dialogues we have about depression is interpreted differently in different times and cultures, meaning humans interpret themselves differently under different circumstances. If we aim to establish universal truths about madness or reason, we neglect the porousness of concepts, and this is not only problematic with regard to understanding these concepts, but also with regard to the position of the mad (or the depressed) in society. For example, the strong administrative emphasis on mental health care, which is expressed in measuring and mapping states of being, or filling in forms, aims to establish a definition of a person that fits diagnostic schemes (Fisher 2014). Talking therapy does more justice to the dialogical aspects of the human condition, but there too we find discourses that rely on problematic categorizations (Fisher 2014).[4]

3 The same applies to the history of philosophy (De Ville 2010).

4 In *The limits of my language* (2021b) I discuss conversations in the context of psychotherapy in detail.

Moving from medical to political conversations about depression

In western societies, depression is currently usually seen as a mental illness, following from low serotonin levels, genes and/or trauma: a disease (and deviation from the norm) that can be solved by taking pills or therapy. Using this capitalist-medical lens to discuss and analyze the phenomenon, however, obscures the political and social structures that affect depression. Mark Fisher (2014) draws attention to the influence of capitalism on humans' self-understanding. He argues that capitalism causes depression through alienation, and frames how humans view depression through propagating a consumerist view of health, portraying the subject as an autonomous person responsible for their own happiness. While low serotonin levels exist, Fisher writes, they do not tell the whole story, and viewing depression as an illness that happens to individuals obscures what is making people ill. Ann Cvetkovich (2012) takes a similar stance and argues that depression is often caused by structural injustice in western societies. She writes that not only capitalism but also racism and other oppressive structures create social conditions that make humans ill.

Shifting from a capitalist-medical to a critical-political lens matters for thinking about the dialogues we have with ourselves and others about depression. Societies should recognize that the dialogues humans can have with themselves are not just formed by their chemical makeup, but by structures of injustice too. These affect how we can think about our position in larger society, which influences how we see ourselves. The dialogues we have with others about depression are not neutral, but shaped by unequal power relations as well – for example, when I was young my depression was not taken seriously by health care professionals, because I was a child and adults did not listen to what I said, which affected the care that I received. Challenging this requires having conversations about structural injustice and mental health at different levels of society – in the medical world, but also in politics and education. These conversations should include how we can reconceptualize care, not just as something that we rely on when we are ill, but as something that should be embedded in all systems, like it is for the mice.

At the same time, depression cannot fully be explained by political, social and economic forces, similar to how thinking about serotonin levels alone is not enough. There is also an existential dimension to what was long called melancholy. We all need to find a way to relate to our mortality and that of others, to the fact that we are alone existentially, and other basic difficulties of life. In the existential conversation about depression, art, philosophy[5] and nature might have a more important role to play than conversations with a therapist.

How to live well with oneself in a violent world

In the existential conversation I have with myself about depression, over the years the focus has shifted from not knowing how to live with myself to not knowing how to live in this world. Even though depression is an isolating experience, the ethical questions that are connected to depression – what it means to be a good person, how to make your life matter – are intrinsically connected to living well with others. This offers clues for how to deal with depression: for me, becoming rooted in life, through activism, my work, living with nonhuman companions, is the best protection against the isolation that depression brings. And when I am not doing well, my animal companions care for me. But this connectedness to the world can also amplify feelings of worthlessness and despair, because those who care for and about other animals are constantly confronted with the large-scale violence against them.

Adorno captures the impossibility of living well in a world that is so violent with the following sentence: 'Es gibt kein richtiges Leben im falschen'[6] (Adorno 1974, 39, as cited in Butler 2012). This means not only that we cannot live a good life when we oppress or exploit others: when the conditions under which we live harm others, our life might be comfortable but we cannot call it good. Judith Butler (2012, 2022) locates two questions in Adorno's remark: how can one live one's life well in a world structured by inequality, oppression, and effacement, and what form does this question take in our time in history? In their discussion, Butler draws attention to how ideology makes certain lives grievable and others not. They emphasize that we are all bodies that stand in relation to other bodies, and need support structures to live instead of simply to survive. Butler further conceptualizes the good life (via Arendt) as a social form of life, in which our lives are always embedded in a larger social web in which we are connected to others who are other than we, and in which many of those others are currently treated violently (2022). Living (instead of surviving) is therefore necessarily connected to

5 Philosophy can help us to better understand the meaning of the experience of depression, and to further investigate what the concept means and could mean in our time. For example, Heidegger's discussion of boredom and Angst (1927) as experiences that ground our mode of being (or Being, in Heidegger's terminology) can shed light on the type of experience depression is. Another example is Camus' connection between absurdity and freedom (2013). Instead of viewing the absurdity of life as something that needs to be repaired or at least changed, Camus writes that it is our starting point as humans, on the basis of which we can choose freedom.

6 Translated by Jephcott as: 'Wrong life cannot be lived rightly.'

resistance and protest, Butler writes, again turning to Adorno's words: 'Indeed, I would almost go so far as to say that, given the way that the world is organized, even the simplest demand for integrity and decency must necessarily lead almost everyone to protest.' (Adorno 2000, 167).

While Butler mentions more-than-human animals in *The Force of Non-Violence* (2022, 41, 199) and acknowledges that they too are part of the fabric of social life into which all of us are woven, they do not seriously consider their precarious position in the webs of relations we form with others. But many nonhuman animals, such as farmed animals or wild animals whose habitats are threatened, are amongst the social groups who are most oppressed, and who have most to lose in the ongoing ecological collapse (Meijer and Blattner 2021). Caring for them, including building new multispecies new life-worlds with them, is a duty for those who benefit from the current system or who partake in it, as well as one of the most radical acts of protest under the conditions of neoliberal anthropocentrism. Furthermore, the most pressing problems for all beings on the planet in this age concern more-than-human phenomena: the climate crisis, loss of biodiversity, extinction of species and other ecological issues. These disasters will make all of us more vulnerable in the time to come, and influence other political and social crises.[7]

Butler is not the only philosopher who does not seriously consider the lives of nonhuman animals, and I discussed different reasons for philosophy's neglect of other animals in earlier chapters (see also Meijer 2019; Derrida 2008). Not thinking about and with other animals is, however, not only violent with regard to them, but also sketches a too narrow view of thinking. And the question of how to live with others in a time of ecological collapse should go hand in hand with the question of how to think with others in a more-than-human world. In the last part of this chapter, I therefore shift the focus from conversations about depression to what it means for philosophy to view thinking as an embodied dialogue with oneself, again in response to Arendt.

Public thinking in the multispecies world

In this chapter I drew on Arendt's ideas about thinking as a dialogue between the I and I to develop an idea of speaking with oneself that is embodied

7 Animalization not only marginalizes nonhuman animals: human groups are also animalized (Ko and Ko 2017, Taylor 2017). Wadiwel (2024) calls this phenomenon 'hierarchical anthropocentrism'. In this ideology, humans are worth more than other animals, but some human groups are more human than others.

and situated, and connected to speaking with others in different ways. I now want to take a closer look at how Arendt links thinking to living well with others, in order to develop a multispecies model for public thinking.

In *Judging*, the unfinished third part of *Life of the Mind*, Arendt connects the dialogue we have with ourselves to living well with others in the polis. She does so through an in-depth reading of Socrates' position on the matter (2021, 614–7). For Socrates, humans were not rational beings, Arendt writes, but thinking beings, and thinking is manifested in speaking (2021, 614–7). Speaking and thinking together form *logos*, and Arendt adds that the dialogue we have with ourselves is the first condition of thinking (2021, 246). This dialogue is politically relevant, because in it we establish the norms that we need in interacting with others. These norms follow from the need to be a friend to ourselves (2021, 246) – 'the self is a sort of friend', Arendt writes about Socrates.[8][9] Not wanting to be a murderer follows from not wanting to be friends with a murderer. This consensus with oneself, which Arendt compares to the Categorical Imperative, is etho-ontologically prior to the one we have with others. So, while the thinking dialogue of I and I takes place in solitude, it is also an integral part of being and living with others (2021, 626). For Socrates, this happens in a situation where both the relation to philosophy and politics are intact. Arendt locates a break in this intactness after Socrates' trial, which led Plato to move away from 'doxa', the way in which Socrates relates to truth as part of the world through the formation of opinion, to a philosophical way of life which is aimed at another kind of

8 The notion of friendship (which Arendt later in *On Humanity in Dark Times: – Thoughts about Lessing* discusses as a political phenomenon) that Arendt posits as a focal point in the relation with oneself and as what provides a foundation for relations with others, is worth exploring further in conceptualizing nonhuman animal inner dialogues. As I discussed with regard to multispecies virtue ethics, treating others as you would want to be treated yourself, is common in dogs. Olli does so by greeting everyone and respecting their boundaries; Doris by respectfully staying very quiet when she visits the house or garden of another person. Their judgments follow from (embodied) reasoning about social situations, based on earlier experiences, their temperament, and so on. How acting towards others as if they are friends relates to their inner dialogue I do not know, and I also do not know if they contemplate in silence, in the way Arendt describes, or prefer to meditate when they are silent and think intuitively when the situation asks for it, but the dialogical model of self can function as a starting point for further exploration of animal selves.

9 One striking point in the human case is that in the dialogues with ourselves in depression, we are not our friends. The most one can hope for in dialogues during this time is to experience some relief by focusing on something else, or accepting that you are not your friend. This also affects relations with others. I do not think you need to love yourself before you can love others, as the cliché says, but some dialogue with yourself is needed for entering dialogues with others truthfully.

truth, outside of the world that humans form, which destroys the plurality of the human condition in oneself. This kind of philosophical thinking is separate from worldly affairs and politics, whereas Socrates' way of dealing with truth corresponded to both.

Building on the insights from earlier chapters about dialogues and agency, my analysis of the role of the conversation with ourselves in living well with others differs from Arendt's on three points. The first concerns porosity. Arendt locates thinking in the individual, and writes that we first need to be a friend to ourselves, before we can be a friend to others. But because we are always already with others, which is constitutive for our way of being in the world, including our normative framework (Donaldson and Kymlicka 2015; see also Rietveld 2008), we cannot say that the discussion with ourselves about norms is ontologically or chronologically prior to that with others. Instead we go back and forth between speaking with ourselves and with others, and this movement begins before humans can phrase their thoughts in the language-game of rational argument. Recent studies about nonhuman animal normativity shows that there are varieties in how we learn social norms (Monso et al. 2022; Vincent et al. 2018). Locating thinking in the mind in this way is also problematic with regard to the role that the body and chemical processes may play in the discussion with ourselves (Bennett 2010, Chapter 3), and the influence of social and other structures on how we may speak with ourselves (Butler 2022).

The second point concerns the image of the human that underlies this view of thinking. Recent Arendt scholarship emphasizes the value of Arendt's work for thinking through ecological questions, and even for thinking about animals and plurality (Vasterling 2021) and animals as political agents (Rossello 2022). However, while Arendt recognizes that our way of being is a social way of being, and her understanding of plurality is a fruitful way of thinking about otherness and difference, she draws strong boundaries between humans and animals based on species-membership when it comes to language and thinking with others. At no point in *The Human Condition* or *Life of the Mind* does she seriously consider dialogue with animals, even though she mentions, for example, that animals appear to others in public and can pretend (2021, 74). Vasterling writes that in *The Human Condition* 'Arendt seems to endorse a human-animal distinction that is exclusionary and homogenizing. Though both are creatures of nature who inhabit the earth, Arendt, in this book, separates humans and animals by attributing world and plurality solely to humans. By excluding animals from plurality and worldliness, Arendt homogenizes animals as undifferentiated and world-less living beings. For this reason, one can argue that the human-animal distinction is dichotomous and essentializes animals' (2021, 16). She argues that Arendt does discuss these notions in relation to animals in *Life of the Mind*. I appreciate Vasterling's discussion of appearing, and her proposal to rethink plurality and worldliness in Arendt, following her recognition that other animals have an 'urge to self-display' (2021, 124). But in *Life of the Mind* Arendt still draws strong distinctions between humans and other animals when it comes to language which affects the possibility for multispecies dialogue, but also for animal thinking, for the dialogues that other animals have with themselves (see Vasterling 126–129 for a discussion; see Meijer 2019 for an alternative understanding of language). This emphasis on species capacities in thinking alone and with others does not do justice to the embodied multispecies character of our worlds, nor to overlaps, relations, and entanglements between humans, other animals, and other beings, nor to human intraspecies differences, nor to the situated character of thinking more generally.

Relatedly, and this is the third point, to analyze Socrates' stance towards thinking, Arendt uses the language-game of the philosopher, by which I mean a specific kind of logical thinking (which uses arguments and relies on human words), to explain what is at stake in thinking, and, more importantly, as a model for thinking and engaging with others. While using the language-game of philosophy to explain what is at stake can be said of all philosophers, including me in this book, it is worth problematizing it when we want to define or even think about 'thinking'. Many humans and probably all nonhuman animals do not think as philosophers do; and as we saw in relation to depression, an individual can also experience different forms of 'rationality' throughout her life.

To be clear, I do not wish to deny that there are no specifically human language-games that are well suited for critical thinking, nor that there is a need for such thinking in life, including when one is depressed, but also for example in scientific endeavors, or in public affairs. I am using the same tools in this book. But I do want to say that this is not the only way of thinking and even critical thinking there is.

Viewing thinking as an embodied and situated dialogue can inform how we do philosophy with others. Thinking with nonhuman – and many human – others, is difficult within existing academic language-games. There are different proposals for changing this in the literature. Before, I discussed Bennett's view about 'writing up' (2020, xx), which she develops as a method of writing that does justice to the influence of nonhuman agency on human thinking, and in the next chapter I discuss Hans Asembaum's (2022a, 2022b) ideas about democratic theorizing. These thinkers however still rely strongly on human language in developing new methods. Embodied and situated thinking with human and more-than-human others is also absent from much of 'public philosophy', which is often understood as a translation of actual philosophy to a general audience, or as self-help (think for example of the School of Life).

Conceptualizing a different kind of public philosophy asks for taking the many forms of language seriously, for example by mapping the genres in which we (can) philosophize, exploring relations between language-games (for example, poetry and philosophy), including nonhuman and human language-games that do not rely on words, listening to others, and developing new practice-based methods of collaborative thinking. The dialogue that a philosopher has with herself is always already part of the world. The dialogues that philosophers have with others could acknowledge and reflect this more strongly, as an act of protest in our time of capitalism, efficiency, and anthropocentrism, but also in recognition of the fact that philosophy always has been a radically worldly practice, in its commitment to dialogue and to critique. In the next chapter I further explore how we can develop new forms of public thinking, together with two groups of human children.

CONVERSATIONS WITH CHILDREN AND ANIMALS

When the toad migration is well underway, I tell the children next door that we are walking again. They usually want to join one or more walks in the weekend, and often bring friends. Some of the volunteers also bring children, and a woman living at the end of the street assists the amphibians coming out of the alleyway next to her house together with her grandson. An older boy who volunteers with his mother is really into salamanders, and invented a ladder to help the frogs to climb out of the gullies – they understand his intervention very well. One woman always brings all her four children on toad rounds, and her eldest daughter sometimes does a round late at night after she went out partying in the weekend. Children are usually very good at spotting the amphibians. Their presence also changes the atmosphere of the neighborhood, and while the amphibians do not care much about their enthusiasm, human adults are affected by it.

'I miss Doris,' R. tells my sister after Doris and I left. R. is three years old, and Doris is scared of children, so when we visit R. and her sister U. who is four, we have to be careful. Doris is always very happy and excited to see them – she whines when we get out of the car and wants to run inside, but once we are inside the house, she prefers some distance. The children really want to touch her. We usually take some time before we approach her so that she can relax, then they give her little scraps of food and can gently stroke her side and back until Doris has enough, which is usually after three strokes or so. They know that if Doris moves away from them, she does not want to be touched anymore, and respect that, even though they would like to continue. They cannot hug her like they hugged Olli, nor can they run up to her freely, but it does not bother them, they love her just as much. Doris is learning to accept their movements and ways of expression, and to relax in their presence.

I often feel that I was raised by horses. When I was a child, I spent all of my free time in the stables. Even though the humans in the horse world were not so great to be around because they were rough with humans and nonhumans, and many treated horses like objects, I felt safe because the horses were there. It is peaceful to be around them. I often felt (and still feel) that horses had a better sense of justice than humans. And of course, they are kinder too. When I grew up my friendships with other animals, such as guinea pigs and cats, were more important than the ones I had with humans, even though I always had enough human friends too.

8. Discussing Multispecies Futures with Children[1][2]

Abstract: I discuss here the importance of multispecies education, which is education aimed at living well together in multispecies communities, in which humans and other animals can be teachers (and other more-than-human beings too, but in this chapter I focus on animals). Multispecies education should not be developed solely by adult humans, but also include the voices of human children and nonhuman animals of different ages, two groups that are usually excluded from decision-making processes in society. Including their voices in knowledge creation matters epistemologically and democratically. To include their voices in this chapter, I discuss multispecies education and the future more generally with two groups of human children, through speaking and drawing.

Keywords: child philosophy, children's rights, multispecies education, multispecies philosophy, democratic theorizing

Learning to speak better with human and nonhuman others is intertwined with learning to live differently with them. Questions about how to co-exist in a multispecies world cannot be answered by humans alone: as I discussed in earlier chapters, this would repeat the anthropocentrism that underlies much of the current violence towards human and nonhuman others, and reinforce the idea that humans know best. Furthermore, as I also discussed before, for normative and epistemic reasons nonhumans should also have a right to speak about questions that concern their lives and futures (Celermajer et al. 2022; Donaldson and Kymlicka 2011; Meijer 2019). In order to understand how more-than-human beings speak, what they say, and what they want,

1 I want to thank Miriam Reeders and the children of 't Sterrenbos, and AnneMarie Tiebosch and the children of Buurtlokaal Aanwezig en Bezig for their participation.

2 This chapter is based on the article 'Developing multispecies education with children and animals' (2023).

Meijer, Eva. *Multispecies Dialogues: Doing Philosophy with Animals, Children, the Sea and Others.* Taylor & Francis Group, 2025
DOI: 10.5117/9789048564415_CH08

humans need to change how they behave towards others, and we all need to learn to live together in new ways (see also Blattner, Donaldson and Wilcox 2020; Donaldson and Kymlicka 2015; jones 2023; Meijer 2021a).

Learning is underexplored in political philosophy and democratic theory, as well as in political practice, where interests and views are often seen as fixed and not as open to transformation. In this chapter I focus on multispecies learning, a topic that I discuss with members of a group that is currently not taken seriously enough in politics or philosophy, namely human children (hereafter 'children'). I also pay attention to the democratic dimensions of knowledge production. With multispecies education I mean education that teaches human and nonhuman animals to live well in multispecies communities: a form of education that offers an alternative to anthropocentric or single-species education, in which both humans and nonhumans can be teachers and students (Tammi et al. 2023).[3] Multispecies education can, for example, play a role in multispecies communication, play, knowledge creation, and work (Editorial Team 2023). What we learn, how we learn, from whom we learn, and where we learn is of fundamental importance in building more just and caring communities and developing the skills that are needed in a multispecies world in times of crisis.

While 'multispecies education' is of importance for humans of all ages, animals, plants, and other beings, in this chapter I focus on nonhuman animals and human children. Children and nonhuman animals are two of the groups most heavily affected by the current ecological crises, and they have no or very little democratic voice in debates about the future of the planet, or other political questions that concern them (Donaldson and Kymlicka, 2018). There are parallels in their democratic position (Donaldson and Kymlicka, 2018) and with regard to their politico-linguistic exclusion (Meijer 2023b). The fact that they are not taken seriously in democratic decision-making is morally and epistemologically problematic: not only should animals and children have a right to speak about questions that concern their lives and futures, they are also capable of doing so and have a unique perspective on life and common affairs (Donaldson and Kymlicka, 2018). Furthermore, in recent years there has been an increasing amount of scholarship arguing that speciesism is a learned ideology (e.g. McGuire et al. 2022; see Donaldson and Kymlicka 2018 for an overview), and if we stop teaching human children to be speciesist, this could drastically change how we live with more-than-human beings.

3 Some authors (notably Acampora 2021) use the term 'zoögogy', but this method of education mostly focuses on educating humans.

Because children have their own perspectives on multispecies education and living well with others, in what follows I not only consider this topic from a theoretical perspective, but also include two conversations with groups of children, one focusing specifically on multispecies education, and the other one focusing on the future more broadly. In the chapter I do not speak with nonhuman animals about multispecies education and the future: as I will discuss in more detail in the conclusion, the next step would be to think with nonhuman animals and children together about these questions.

Education in multispecies communities

In recent years, there has been much attention for the multispecies community in animal philosophy (e.g. Celermajer et al. 2020; Donaldson and Kymlicka, 2011; Haraway 2008) and in critical scholarship about children (e.g. Editorial Team 2023; Roussel 2023; Taylor and Pacini-Ketchabaw 2018; Tammi et al. 2023). The starting point of both of these strands of scholarship is that our communities, cities, and nations are and always have been multispecies, and include a variety of relations between humans and other animals, that range from avoiding each other when sharing habitats, to sharing beds and sofas in the case of companions. In these relations, both human and nonhuman animals exercise agency in various ways. Theorizing the multispecies community is a critical project which includes analyzing relations of domination (Wadiwel 2015). But animal philosophers also focus on phrasing normative obligations towards other animals, and how to achieve interspecies and multispecies justice (e.g. Celermajer et al. 2020; Celermajer et al. 2022; Donaldson and Kymlicka, 2011; Emel and Nirmal 2021; Tschakert et al. 2021).

In working towards more sustainable, just, and caring multispecies communities, education can play different roles. Humans need to educate themselves by making space for and listening to other animals – we should 'unlearn' anthropocentrism, in ways similar to unlearning racism (Srinivasan 2022). This is not just a political, but also a cultural project, to which this book aims to contribute (see also Hertbrechter 2023; Moran 2022). But in addition to unlearning violent ways of being, animals of different species, including humans, also need to learn to interact better in order to share the world in new ways.

Importantly, in learning to co-exist in better ways, nonhuman animals can be teachers too. There are many ways in which humans can learn from and with other animals.[4] I will here mention four, but this is by no means meant as an exhaustive list. First, humans need to learn from other animals about what kinds of beings they are. Centuries of oppression have distorted views we have of other animals, in science (Despret 2016), culture, and politics (Donaldson and Kymlicka 2011); countering this asks for new forms of engagement with them. Second, humans should learn from and with other animals how to better live together. I do not mean how to take care of more-than-human animals in the right ways, but rather rethinking 'community' together, using political concepts to guide us (Blattner et al. 2020; Meijer 2022). Through embodied experiments and practice-based research, humans and other animals can collectively learn how to co-govern shared communities. These experiments already take place in certain animal sanctuaries (Blattner et al. 2020; jones 2023). Third, human societies can learn from animal communities about how to live better on the planet with regard to natural resources. This is recognized in approaches to nature conservation that take into account animal cultures. Nonhuman animal ways of life are generally more sustainable than human ways of living, at least those in rich countries, and learning from animal ways of being can guide us in protecting natural areas (Bell Rizzolo and Bradshaw 2019). Fourth, we can learn from other animals about the meaning of life and how to live well. The basics of existence – love, loss, and community – are the same for many animals. For example, from Olli and the mice I learned about care, courage, kindness, dealing with illness and death, and other things that matter in life.

In order to better understand the role of animals as teachers, it is worthwhile to look outside of the western tradition of thought. In Daoism, and specifically the Zhuang Zi, animals and natural entities are often described as beings to learn from, because they know how to live in line with *dao* (the way of life). Other more-than-human beings can be teachers too: Kimmerer (2013) for example mentions that plants are seen as teachers by different indigenous communities, because they have been on the planet longer than humans and exist in other ways than humans do.

There seems to be a large gap between learning from and with other animals and the current everyday meaning of 'education'. However, there are already examples of multispecies education, such as dogs who visit primary schools in order to teach children about how to communicate with them,

4 The ideas of the main figures in the philosophical canon were shaped by their relations with nonhuman animals, Kelly Oliver (2019) argues, leading her to conclude that other animals taught us to be human.

and the training of both humans and dogs in certain dog schools (Meijer 2020). There is also an increasing amount of literature about the social lives of nonhuman animals including their normative behaviors that can help guide human and multispecies learning. Furthermore, the terms 'education' and 'learning' have different meanings in different (human and nonhuman) communities and cultures, and refer to a wide variety of practices, some of which are local, or species-specific, while others are more general (like: do not bite others randomly).

Existing forms of education can function as a starting point for multispecies education. For example, in Scandinavian countries there is a tradition of nature education, which includes lessons outside, and practice-based ways of exploring the natural world. These can be extended to include new forms of interaction with nonhuman animals, based on consent. However, learning with and from other animals has not been something we have taken seriously as societies, so we should also develop new educational practices, in households, cities, and on farms. The perspectives of animals should be foregrounded in these investigations, and those of human children too. Similar to animals, children are also generally not consulted when adults develop regulations for education, nor to discuss the content of education, even though they have a unique perspective on questions about education, and what they learn and how they learn greatly affects their daily lives, as well as their future.

The right to co-shaping education for animals and children

Nonhuman animals and children have their own perspective on life, the good life, and how and what they want to learn. They also have most to lose, in the context of the climate crisis. Therefore it is not enough to develop new educational programs in a multispecies context without them: they should have the right to co-author these. However, under current circumstances both groups are excluded from deliberation about education and from politics more generally, because adults assume that they are not capable of democratic engagement, or that they would not benefit from participating in deliberation about political questions that concern them (Wall 2016).[5]

5 Donaldson and Kymlicka (2018) write that the political exclusion of animals and children shares similarities, which follows from a specific rationalist social contract view of democracy, which they call the 'capacity contract' model. This model is unnecessarily exclusionary, they argue, and does not take all our intuitions with regard to politics into account. In order to take

In my discussion of deliberation with mice, I wrote about already existing political experiments with more-than-human animals, that belie the idea that they are not capable of democratic action. Similarly, there are many examples of projects that involve children in political decision-making and show they want to and are capable of expressing themselves democratically (Wall 2016; Wall and Dar 2011). In many countries around the world there are examples of children's parliaments, children's councils, and child mayors (Wall and Dar 2011). As Wall and Dar (2011) show, these projects do not automatically give children more voice: their role is generally advisory and sometimes their presence is merely ornamental. Furthermore, if children are consulted within already existing political frameworks, adults will still determine how much their voices weigh, even though their political participation may require other political arrangements, either because of their capacities, or because their interests are different than those of adults, as is the case with the climate crisis (that would require more future proof forms of political decision-making). In thinking about developing new forms of political communication with children and new political models, the movement for children's rights can learn from recent developments in animal philosophy, where embodied and situated encounters, dialogues, forms of deliberation, co-government and related practices are conceptualized politically and socially, together with animals of different species (Donaldson and Kymlicka 2018).

Democratic theorizing with children and animals

Engaging with animals and children when developing new forms of multispecies education is not just important in political practice, but also in creating knowledge. Knowledge creation is currently in the hands of a select group of adult humans. As Hans Asenbaum (2022a) shows, this is problematic from the perspective of democracy: when you write about inclusion, agency, and transparency in relation to marginalized groups it is paradoxical not to include their voices and perspectives in developing these concepts. Asenbaum writes that theorizing currently takes place the 'academic ivory tower' (2022a, 1), by humans who were trained in specific educative institutions, which requires social and economic resources not available to everyone. This affects how, and what kind of, knowledge comes into

seriously not just the interests but also the perspectives of animals and children, they propose another model of citizenship, which is based on membership, and which recognizes that our societies are already multispecies. Donaldson and Kymlicka focus on domesticated animals' right to co-shape the shared multispecies community. I would add that liminal and wild animals also should have a voice in these matters, for example when we share habitats.

being. Asenbaum draws attention to how underlying processes of systemic inequality, rooted in capitalism, heteronormativity, and colonialism – and, I would add, anthropocentrism and speciesism (Acampora 2021) – inform our ideas about what counts as knowledge. Drawing on critical indigenous grounded theory, participatory research, and assemblage theory, Asenbaum argues we need 'democratic theorizing' instead of 'theorizing democracy'. Democratic theorizing includes taking seriously marginalized knowledges in theorizing, for example through dialogues.

I share Asenbaum's concerns about theorizing about others without engaging with these others, and his trust in dialogue, broadly construed, as a way of developing new methodologies. In the case of nonhuman animals, theorizing about them instead of with them often reinforces silencing them (Meijer 2019). As I will discuss in more detail in the next chapter, listening is an important part of democratic theorizing with more-than-human beings, and children too (Lamberti 2023).

Learning to think with children and animals, and to collectively create new forms of knowledge, needs time. Asenbaum (2022b) discusses the importance of taking the process seriously, instead of moments of decision-making, in relation to his collaboration with Black Lives Matter activists. In order to avoid knowledge colonialism, taking and appropriating insights from marginalized groups (human or nonhuman), it matters to build a relationship and offer something in return. However, this kind of engagement is also labor-intensive, and can be burdensome for both scholars and participants. With animal companions, this tension can be resolved because day-to-day interactions can function as the dialogue between researcher and participants. But this is not always the case, and we therefore also need to think about other methods to create knowledge with more-than-human beings and children.

One method could be the assembly (Meijer forthcoming b). In deliberative theory and activism, citizens' assemblies are currently promoted as a way of improving democratic decision-making. For example, on their website environmental activist group Extinction Rebellion argues for the implementation of citizens' assemblies in political decision-making regarding the climate crisis. In citizens' assemblies, randomly selected adults meet to have informed conversations about a topic, and present their decision, which can be binding or advisory, to local or national governments. This political model can also help us think about setting up dialogues with children and, in a modified form in which spatial arrangements facilitate the discussion, nonhuman animals and other more-than-human beings (Meijer forthcoming b).

Setting up assemblies to discuss multispecies education with children and nonhuman animals exceeds the scope of this book. But as a first step towards including the voices of children in writing about multispecies education, I discussed this topic with two groups of children. I should note that my conversations with the children are real conversations, and not interviews as a research method to collect data. In the conversations, I mostly asked about their opinions and ideas, but also discussed mine, and I answered the questions that they had for me. While I had written down questions and themes to discuss beforehand, I mostly improvised during the conversations, using the Socratic method of conversation that is often employed when philosophizing with children.

Speaking with children about children's rights, education, and animals

On January 11, 2023, I visited a class of 24 11- to 13-year-old children at Ssbo 't Sterrenbos, a special educational needs school in the middle of the Netherlands, to discuss multispecies education.[6] We spoke for around fifty minutes and were seated in a circle. Their two teachers and one teaching assistant were also present, and participated in the conversation. I had prepared questions, but the conversation flowed naturally around the topics that I wanted to discuss with the children, so I left my notebook closed. The conversation addressed four themes: children's rights and education, children's political voice and participation, animal rights and voice, and poverty and work. I will not describe our conversation about poverty and work here, because it is not related to the topic of this book, but it was clearly something the children felt strongly about.

I was the first philosopher that the class met, and after a round of introductions in which they mentioned their name, which companion animals they lived with, and their favorite animals, we spoke about philosophy, and what it is that philosophers do. Philosophers think, was the consensus, and one girl mentioned that they think about the future. Another girl mentioned that philosophers think about human rights, and this immediately steered the discussion in the direction of children's rights.

We first discussed which rights are important for children – they mentioned privacy, playing, and learning. All children felt strongly about what they learn in school. Reading and learning languages were mentioned as

6 For reasons of privacy, I will not mention their names in this text.

important, and understanding the news as well. Topics that they wanted to learn more about were history, foreign languages like Chinese, and traffic. I asked the children if it was important to them to be able to determine what they learned in school and while most children thought that was the case, some had doubts. Sometimes adults know more about a certain topic so are in a better position to judge what they should learn, a few children thought, and one girl mentioned that she sometimes disliked certain topics and therefore thought they were unimportant, while they were in fact important. Their arguments in favor of participating in thinking about their education were that they know best what is important for children since they are children, and that adults do not always make the right choices. The children mentioned that they had the feeling their voices were heard, in school and by their parents.

Having a voice was also seen as important by the children in other contexts, at home and in politics. Specific political topics they mentioned were the climate crisis and poverty. The children knew that there are possibilities for democratic participation for children, such as the children's question hour in the Dutch parliament, and children's councils. They suggested that their political participation could be improved by better access to politicians, and that politicians should actually follow their suggestions, and not just listen to what they say and go on with their business. This led us to discuss listening, and the different forms of listening there are: children often need to listen in the form of obeying, but for a conversation, such as the one we were having, you need another kind of listening, in which you focus on understanding the other.

For reasons of time, I changed the subject to animals. We had been speaking about different kinds of rights so I asked the children if animals also should have rights. There was a very loud yes from the group. However, some children did object that they do not need the same rights as humans; these children wanted to improve animal welfare, while others held a strong animal rights position. The latter prevailed, and arguments for animal rights were that they have feelings just like us, that they are like us, and that pigs and others in intensive farming often do not see the outside world in their entire lives while humans can eat fake meat that tastes just like meat. The children also held strong objections to zoos, and specifically mentioned it should be forbidden to take animals away from their homes and families. Only when animals are severely injured, that could perhaps be a reason to care for them in a zoo. They also felt sorry for wild animals in circuses and were against the use of all animals in circuses.

I asked what they thought about listening to animals, and whether that matters for animals like it does for human children. Most children thought so, because the animals have thoughts. They can express themselves in different ways, with their voices and faces, but also by making nests, and dogs sometimes make a nest of blankets on the couch. A way to listen better to the other animals as societies is to consult humans who study their feelings and thoughts, the children mentioned. I asked about multispecies learning. Young animals also need to learn, from their parents, or from humans when they live with humans, but the idea of learning together with animals in the school building was immediately rejected, since this would make the animals uncomfortable. Yet there were many things they wanted to learn about animals, for example about the language of brown bears.

We then spoke about what it means to lead a good life, about the strangeness of life, and about work, money and poverty. At the end of the conversation, I asked the children if they had questions or if there were things that they wanted to say. There were a couple of questions about my work, and then the children returned to the topic of the future. Adults can decide, but it is our future, said one boy. Another boy added: we are the ones who have to live longer on this planet. You don't know that, said another boy, maybe you die tomorrow, but the point was clear.

Not all children in this group could respond immediately to questions. Some children needed more time to think about responses, had trouble speaking, or could not participate actively in this conversation for other reasons. All children however paid attention and engaged enthusiastically. Some expressed their approval in nonverbal ways, for example by nodding. Others contributed to the discussion by making sounds that expressed approval or disapproval. While our conversation raised more questions than we could answer, it was obvious that the children had a lot to say on the topics we discussed and were eager to share their perspectives. They also expressed a desire to continue with philosophy. Possible further steps they mentioned are creating spaces where encounters with animals can take place, writing letters to animals to formulate their own questions, and creating a collage together with drawings and text, to think further as a group.

Some of the themes that came up in our conversation can also provide a starting point for developing new methods of multispecies education. The first concerns space. At the end of the conversation the children mentioned that they would prefer to learn about and with other animals not in school but in a playground or at least an outside space, with opportunities for different activities. I agree that multispecies education should begin outside. While invading others' habitats is to be avoided, many natural spaces, parks and urban areas are already shared between animals of different species.

One way of engaging with these animals is through play. The children told me play was an important activity for them, and play is important for other animals too. Playing together can help with learning about the other, setting boundaries, and improving overall joy (Irvine, 2001). And finally, according to the children, both they and the animals should have a say in questions that concern their lives, including education. The children emphasized that this asks for developing institutional mechanisms that they can affect, for actual democratic power, and that the same is true for other animals. After all, it is their future too.

Drawing the future with children in Buurtlokaal Aanwezig en Bezig

On May 20, 2023, I conducted a second conversation with children about animals and the future, this time in 'Buurtlokaal Aanwezig en Bezig' in Amsterdam. The children present usually visit the Buurtlokaal, a community space in the Bijlmer neighborhood, on Saturday afternoon between 2 and 5 pm. On these afternoons, they talk, make drawings, play games, drink tea and eat a sandwich. They visit the Buurtlokaal because their parents are working or not capable of taking care of them for other reasons. On this afternoon, ten children were present, in ages ranging from six to twelve. Later in the afternoon, other people came to visit too, mostly elderly women.

When I arrived, the children and I sat down around the table and drank lemonade. AnneMarie, who runs the Buurtlokaal and whom the children know well, was also present, and my sister, who lives in the neighborhood, joined us too. After a round of introductions we spoke about philosophy, and discussed the kind of topics that are important to think about philosophically. Similar to the children in the first group, these children had no experience with philosophy, but many questions. The first topic that came up was oxygen, which connects humans and trees, and everybody needs. Food was the second topic they thought to be important, because what we eat and if there is enough to eat matters a lot to humans and other animals. One girl raised the question of whether something can come into being out of nothing, or that there never was nothing, but then, how did the universe originate? Money was also seen as important: we discussed the fact that it does not exist but still has a meaning, and we spoke about the injustice of poverty too. The children also thought that education was an important point for philosophy: the school that you go to, and if that is a good school, matters a lot for your future.

This insight led us to discussing how we can learn to live well together with those who are other than us, now and in the future. We first spoke about what kind of attitude is needed to live together. According to the children, what is most important is to be kind to one another. They all agreed on this as a basic principle. In order to be kind, we do not have to be the same. When you are not the same, you can learn from someone. Sometimes learning happens through arguing, and therefore having an argument can be a good thing, but you should not become violent. One boy raised a problem about rules. If there are rules that people do not want to obey, then it is better to have no rules. Other children mentioned that rules are important and that we need to make them together. The first boy was not sure: he thought that someone in charge should do this, because he would not know how to do it. Maybe the king. But the other children rejected this, and thought that as children they have a specific perspective on rules that matters too. Nobody knows everything, not even the king. The first boy was not convinced. Later in the conversation he also mentioned that adults in general know much more than he does, because he has ADHD.

According to the children, living well with others entails more than being kind. It also has practical implications. They felt very strongly about plastic. Living well with animals, trees, and other beings of nature means that we should not throw our plastic away on the street. The fishes don't know that it is unhealthy to eat and they may die from it. This insight about eating led another girl to say that we should feed the animals who are hungry, for example those who live outside, and everyone agreed about this.

We then moved to discussing how we can live better with others in the future. The first point that the children mentioned was that everybody is equal. In making decisions about the future, everybody counts. The next point was that helping others is important. One suggestion that the children had was to work together with other countries to alleviate poverty. All children present were of Surinamese heritage, except two girls who moved to the Netherlands from Nigeria two years ago, and the children mentioned the relation between the Netherlands and Suriname specifically. Helping others however should not be limited to humans. We should also help the other animals when we can. And we need to keep the world clean: we need to fly less, and close polluting factories. At the end of the conversation, one girl connected the idea of helping others to equality. She said that we need to help one another because when there is a crisis like a war, we are all equal.

Before we began speaking with each other, we had discussed the set-up of the day. The children had told me that they wanted to speak about the future, but also wanted to make drawings. So, after a short break, they began to draw. One boy who did not speak but ate many cookies when the others spoke, made this drawing of trees bearing fruit, which showed that he had followed our conversation closely.

His brother was the boy who was skeptical about his own knowledge compared to that of the adults, developed the following image for the future. Only one house, and a lot of land for the other animals. There are also two dogs in the picture, because he would like a dog companion, and dogs like to be with other dogs.

The girl who raised the question of how something can come out of nothing drew this half-cat, half-girl, as an image for the future.

The most comprehensive vision of the city of the future was drawn by one of the girls from Nigeria, who wants to become a doctor.

After the children finished their drawings, they ate a sandwich and went outside to play. Some came back inside to tell me that they enjoyed the combination of talking and drawing, and two of them thanked me for listening to them.

Because the children enjoyed thinking about nature and animals, we spoke about meeting again in the summer, maybe to investigate which animals live in their neighborhood, in order to find out how to live well with them. The drawings they made complemented the conversation (all children made drawings), because they enabled the children who cannot or do not want to speak out in a group, or who are shy, to participate, and because they offered another mode of expression than words. Similar to the children from 't Sterrenbos, these children were eager to share their views, and capable of doing so.

Further developing multispecies education with children and animals

In this chapter I spoke with two very different groups of underprivileged children, and our conversations gave me many new ideas for research, regarding the form and the content of dialogues with children, developing new forms of education, and children's democratic participation. The children themselves contributed to thinking about moving forward. For example, in the context of multispecies education, the second group and I discussed making a large map of the habitats of nonhuman animals in the neighborhood, which can function as a starting point for learning about them, as a first step towards living differently. This would combine walking, drawing, and thinking on our side, and perhaps engagement with the animals we encounter, if they are up for it.

More generally, further developing multispecies education requires being specific: what kind of education are we speaking about, with which animals, and for whom? In answering these questions with children, existing educational practices and institutions can play a role, if they are extended to include their voices – for example by having a roundtable discussion at the beginning of the week in schools in which children can share their views about education for that week. But in order to really do justice to the perspectives of children more radical steps should also be taken, following the emphasis on play, space and voice I discussed above. Developing new multispecies educational practices requires active engagement with nonhuman animals, so they can co-determine what 'education' means by showing what they want to learn and what kind of wisdom they want to share (Blattner et al. 2020; jones 2023). This means we need to develop methods of practice-based learning, in which different animals of different ages can affect not only the outcomes but also the methods of learning (Tammi et al. 2023). In addition to speaking and drawing, developing educational practices can involve diverse activities, such as walking and other forms of movement, playing music, sitting together in silence, developing common habits, and play (Acampora 2021).

In addition to these experiments, which require long term efforts, assembly type meetings can play a role in incorporating the voices of children (Wall 2016; Wall and Dar, 2011) and animals (Donaldson 2020; Meijer forthcoming b) in democratic processes regarding education or other questions that matter for these groups. The conversations I had differ from citizens' assemblies in several ways. The children do not represent all different groups in society, they did not receive information beforehand, and we did not enter an extended process of multiple conversations in which they made up their mind. I found the first group of children because my friend M. is their teacher, and the second group because my sister lives close to the Buurtlokaal and gave workshops there before. Still, our conversations were fruitful.

The advantage of our informal meetings was that the children felt at ease in their spaces, also because there were adults present who knew them and whom they trusted. This especially mattered for the children in the first group because they are neuro-diverse and some of them express themselves differently from the norm. While I did not build up a relationship of trust with these children, they trusted the situation and the other adults present could assist where necessary.

So, even though the children in both groups are vulnerable in their own ways, in the conversations that we had, many of them were eloquent and passionate about democratic participation, and I would expect them to do well in the setting of a citizens' assembly, or a similar model for democratic participation, if these would take place in a setting where they feel comfortable, and if non-verbal ways of expression would be included in decision-making processes.[7]

Conversations with children are often depoliticized, and in public discourse 'politics' is often equated with strategic power play (Latour 2018; Meijer 2022c). Apart from the normative reasons I discussed above for

7 In thinking about children's participation, space, voice and influence are not enough, the public, who is listening, also matters (Lundy 2007).

speaking with children about political topics such as multispecies education and living well with others in the future, taking them seriously also opens up what we consider to be 'politics' and 'political discourse'. This can be beneficial for human adults too, because it changes the tone of the debate, and brings into view other aspects of the topics that are discussed. An example of how this can work is the yearly *Kindervragenuurtje*, Children's Question Hour, in Dutch parliament in December. During this hour, in which children get to ask questions about topics they care about (ranging from financing for the animal ambulance to sea level rising), the politicians step out of their agonistic role, and become more gentle and serious than they usually are. Most of them answer honestly instead of trying to refute the question or using it to promote their own views, which is common in adult political debates (see also Moore 2019). Taking children seriously can also bring joy and wonder into the discourse. Similar to how this works in the case of nonhuman animals, however, taking the voices of children seriously does require an effort on the side of human adults. It asks us to practice paying attention, and not just to learn to speak in better ways, but also to listen.

LEARNING TO RESPOND

When I see an insect drowning, I try to let them grasp a blade of straw or stick with their arms. I have learned that their wings can get stuck to the straw or stick if I scoop them out of the water, sometimes causing them to tear. Sometimes there is no other possibility than to lift them out, and then I blow them dry and hope for the best.

My father was afraid of death. When he was in the hospital and things were not looking good, he and I spoke about him going home, even though we both knew that the chances of going home were small, because this was what he wanted.

The first pigeon I found on the street had a large wound on her throat. I took her to the bird shelter, where they told me that she would have to be euthanized. In the decades that followed, I found hundreds of birds who were ill or injured on the streets. I learned to recognize sick birds from afar, from their posture, and the least stressful ways of picking them up. No matter how many I had in my hands, touching a bird is always special.

9. Learning to Listen

Abstract: Finally, I discuss how we can learn to have better conversations with human and more-than-human others through an exploration of the different roles of silence, and the importance of listening, in multispecies politics. While certain forms of (non-invasive) ethological and biological research, as well as artistic experiment, can play a role in improving interactions with nonhumans such as animals and plants, the most important task for humans in our age is to take a step back and listen. We should invite others to speak out in existing debates and deliberation about our common life-worlds and habitats, but also let them determine if they want to speak with us at all, and be responsive to their resistance and refusal, as well as possible invitation. Learning to speak better with others – human or nonhuman – begins with listening. In the case of individuals, but also for societies, learning to listen better is not just an ethical but also a political task.

Keywords: dog philosophy, political silences, listening, democratic listening, multispecies community, refusal, animal resistance

I began writing this book in 2023, when I had long covid and needed something to cheer me up. When I fell ill with covid I had never been sick for long, and thought I'd recover soon, but it took me around twenty months. During this time Olli was also very ill, and died. But we had someone who looked after us and protected us during these days: Doris.

Like Olli, Doris is originally from Romania, where she was found in a trash can with two of her siblings. She came to the Netherlands as a puppy, over eight years ago, and had lived in several households before she came to live with me. She had developed what humans called 'fear aggression' and now call 'reactive'. Over the years, we created a strong common framework of language that helps us navigate social situations that are difficult for her. We took many courses at the dog school, because Doris likes the exercises. She befriended some of my dog and human friends, and made her own dog

Meijer, Eva. *Multispecies Dialogues: Doing Philosophy with Animals, Children, the Sea and Others*. Taylor & Francis Group, 2025
DOI: 10.5117/9789048564415_CH09

friends in our neighborhood. But in contrast to Olli, who enjoyed contact with others, Doris never became fond of meeting new humans and dogs, and she is very protective of us and the house. When we lived in Amsterdam, this made her tense: there was constantly so much going on that she needed to keep an eye on. Now that we live in a small village, it is more doable, but still, it's a fulltime job. I never asked Doris to protect me, but I have learned to appreciate it. I know that it matters greatly to her to do the right thing for me. Her vigilance makes me feel safe, especially during the months in which we slept with the back door of the house open because Olli needed to go out at night. Doris escorted Olli on all his trips to the garden, in his last months. For her, responsibility and responsiveness are fully entangled. She is completely attuned to me, and was fully attuned to Olli too. Yet our relationship has also been difficult, mostly because the society we live in does not appreciate dogs who need space. While I always wholeheartedly supported her, I also needed time to learn to respond to her. I had to learn the balance between wanting to keep her safe and supporting her own decision-making.

This example points to a larger problem in our societies which is that we do not have many narratives about how to respond to nonhuman others (and in certain cases human others too, as I learned from being sick with long covid), which makes it difficult to have conversations with them. In this last chapter I do not want to (nor can) define 'dialogue' or give a recipe for having better dialogues, but turn to the question of how humans can learn to respond better to others – as individuals, but especially as societies. I will focus specifically on silence and listening, because behind the word 'silence' we find different kinds of practices, power structures and possibilities that influence the conversations that we have, and can have with human and more-than-human beings. And because as humans we tend to speak much about others, but do not listen enough to them.

From silence to silences

In describing dialogues, I have focused mainly on what has been said, and how it was said: on questions and responses. I emphasized the embodied and situated dimensions of speaking with others, taking seriously different kinds of agencies and languages, and the importance of willingness to understand those who are different from us. But dialogues also involve listening, and taking turns speaking. In other words, they involve silence. Silence is a necessary ingredient of every conversation, including that with yourself, and it is the basis of knowing how to respond. Silence is necessary for listening, making space for the other to speak, and for connecting the dialogue with your self to the dialogues you have with others. This may seem especially important in speaking with beings with whom we usually do not speak, but it matters just as much with regard to those who are very familiar – silence can create space for the new in all relations. Through being silent you can learn to read silences. Learning to read silences is connected to learning to respond better, and to knowing whether you should respond at all, or let the other be.

Silence is not one phenomenon, nor the negative of language, but can take on many forms in politics and public life.[1] Elsewhere (Meijer 2022d) I distinguished between four families of political silences. The first is silence following from exclusion, such as silencing. This can take place in debates (Young 2002), but communities can also be silenced, for example through colonial oppression that marginalizes the use of whole languages (Spivak 1988), and species can be silenced through anthropocentrism (Tuvel 2014). The second form of political silence follows from the shape of the discourse. For example, neoliberal political discourse makes it difficult to adequately speak about values that do not fit into a capitalist understanding of value. The third family of political silence concerns resistance. Silence as resistance includes sit-ins, but also refusal, as I will discuss in more detail below. Finally, silence also offers a space for transformation. Listening and keeping silent make possible new relations. This ordering does not capture all silences – in fact, one feature of silence is that it often escapes us – but it provides a starting point for learning to read silences better in multispecies conversations, and to develop social and political ways forward.

Silencing plays an important role in the conversations we have (and do not have) with nonhuman animals, and plants too. Many nonhuman agents are considered to be silent because of how borders are drawn around the political community and language (Meijer 2019; Tuvel 2019; Wadiwel 2015). Their silence follows not from their capacities, but from anthropocentrism in culture, politics, and other structures that frame and govern relations. This presents a false picture of them, of language, and of politics (Meijer 2019). It also obstructs social progress, because this requires engaging differently

1 The role of silence in politics differs between human cultures. There is a strong focus on language in European and North-American politics, but in Japanese politics, for example, silence plays a central role in determining one's attitude and discourse, following its importance in culture (Lebra 1987; Hasegawa and Gudykunst 1998).

with nonhuman beings. Viewing existing dialogues as dialogues can be a first step in challenging silencing, and acting otherwise. What precisely dialogues entail differs from group to group, but critically analyzing power structures in society and research should always be part of claims about earth others like nonhuman animals and plants, and language, and to build new relations we need to acknowledge past wrongs (Palmer 2012). With regard to the sea or art, other questions are at stake, and the line between speaking and being silent is less clear, because speaking takes on another form and silence does too. Still, new conversations can only take place when we take into account the influence of power relations on how the lines between language and silence are drawn.

Tracking the silences that follow from how we speak with one another, the second family of political silences we can encounter, asks for attending to how ideology structures debates. For example, in capitalist societies many conversations center around use and profit. This enters our language in the words we use – think of the rise of the word 'consumer' (Fisher 2014) in relation to art, education or health care, or the 'production' of knowledge – and it leaves holes in what can be said. A capitalist discourse around art makes it difficult to express value that cannot be monetized; a capitalist discourse around depression, which portrays humans in need of care as consumers, limits options for treatment and obscures the fact that social structures make some people ill; a capitalist discourse around nonhuman animals obscures their agency and subjectivity. Another example is our strong focus on the human. Anthropocentrism invades many conversations about the value of more-than-human beings, ecological problems such as the climate crisis, and the future of planet earth (Celermajer et al. 2020; Youatt 2017; Moore 2016) – with regard to how conversations are held, but also linguistically, for example when 'nature' and 'human' are opposed, or 'animal' and 'human'. Viewing humans as the center of the universe obscures the agency and interests of other beings, and our interconnectedness in the larger scheme of life. Paying attention to what others say can challenge this. For example, elephants have a word for human that also means danger (Soltis et al. 2014).

Further developing new vocabularies with others is necessary to bridge the gaps that follow from how we speak. As I will discuss in more detail below, this begins with listening, which is connected to transformation and possibility. In the space between question and answer, change is possible. But before I turn to listening, we need to take a closer look at silence as resistance, the third family of silences, which is especially important with regard to thinking about (dialogues with) nonhuman animals.

Silence as refusal and respecting 'no'

By focusing on dialogues as a model for reforming relations I do not mean to imply that everyone should engage in dialogue with others all the time. Similarly, in my emphasis on the importance of building common multispecies life-worlds with others, I do not want to say that everyone should be part of all life-worlds. This is impossible and undesirable. I do want to emphasize that we already are in dialogue with many others – human and nonhuman – in ways that we are insufficiently aware of, and that we should acknowledge the agency of earth others in existing dialogues. Most multispecies conversations are currently overshadowed by human domination. A crucial part of engaging more respectfully in conversations with nonhuman others is learning to read their silence, and understanding and obeying their 'no'.

Many nonhumans resist through silence (Hribal 2011; Wadiwel 2015) and much of which we currently read as consent or assent, or do not perceive at all, is in fact resistance or refusal. Doris constantly reminds me of the importance of respecting the space of other animals, and the human failure to do so. While she enjoys dialogues with me, Olli, and her chosen friends, she does not want to engage with strangers in the larger community, and would, for example, not want to be part of forms of deliberation that include physical proximity to strangers, eye contact, loud noises, and sudden movements, or other things that happen when humans and domesticated dogs are together and get excited. She voices her dislike by avoiding these encounters, and resisting when she cannot avoid them. Doris is not the only one who does not want to engage in the dialogues that are expected of them: many other animal (including human) individuals and communities feel the same because of trauma, because they do not like certain activities or communities, or members of certain species, or for myriad other reasons. In our human-dominated world, for many animals and other beings it is difficult to escape interaction with humans, because they forced to live in captivity, or because humans occupy their territories on land, in the water, and in the air. This may lead them to resist, in direct encounters, but also, and this is a common response of nonhuman animals, to refuse by turning away from humans (jones 2009). (Of course, there are also many animals who do not have the opportunity to struggle, fight, or leave the situation, and who simply give up and wait.)

Refusal is also a common human response to domination. Leanne Betasamosake Simpson discusses human indigenous refusal as a response to settler colonialism (2017, 34). She tells the story of Kiizhigo, a Michi Saagiig Nishnaabeg who lived in Curve Lake on Turtle Island (also known as Canada).

Kiizhigo was tired of the settler government interfering in his life and that of his community, so he left to live on an island. With his skills and knowledge of the land, he could take care of himself, and he lived on this island, which is now named after him, until he died. Simpson writes that his refusal is written into the land, and that those who paddle by the island think of his refusal of colonial domination. Kiizhigo's refusal ends a conversation he never asked for. Settler colonialists do not just take part in a conversation, they force the shape of that conversation (including the language in which it takes place) upon indigenous communities. Simpson points out that refusal not only contests existing violent conditions, but also the epistemic order that underlies them and was imposed on those who refuse.

Bonnie Honig (2021) frames refusal differently. Through reading *The Bacchae*, Euripides's fifth-century tragedy, Honig develops a feminist theory of refusal that connects withdrawal to transformation. She uses the figure of the arc to illustrate this. When the conditions under which one lives are unjust, one can move away from the polis, try out new forms of living, and return with that new way of life to the city, in order to transform society, and perhaps take up space in government. Honig's underlying view is that we cannot leave social relations; we are always part of social structures with others with whom we may have very difficult relations or by whom we are oppressed, and should deal with that. Judith Butler articulates a similar point in their discussion of precarity and ethical demands (2022). There is no real outside to politics, Butler writes. We can resist, by coming together with others, and question hierarchies, but this always takes place in a social world that we never chose and can never escape. As Butler recognizes, resisting may be more dangerous for some than for others. In *The Bacchae*, it does not end well.

The figure of the arc that Honig describes resembles that of the conversation. Something is being said by society, part of the community responds by turning away, deliberating, and returning to answer. The society will then respond again – with violence, struggle, or acceptance. In contrast, Simpson's description of refusal ends a conversation. But as the story of Kiizhigo shows, the ending of one conversation opens up others: with the land (in ways that are likely impossible under the conditions of the colonial government), and with one's own community. What may look like silence or withdrawal to those in power, or those who are part of the status quo, can be a commitment to life for others.

According to Honig, refusal creates room for building new types of relation and community. Returning to the polis might change society for all, as social justice movements aim to do, but being able to return to the polis is not possible or desirable for everyone. Nonhuman animal refusal can illustrate this point. For most nonhuman animals – farmed animals, lab animals, wild and liminal animals – living with humans is unsafe, and they cannot frame their demands in the language of power. When they are lucky enough to be able to escape, most of them have no other option than staying away, hopefully finding a way to survive elsewhere. For animals, leaving a dangerous situation or a conflict is often a much better option than staying and resisting, because humans will overpower (and usually kill) them. Nonhuman animals belong to the groups living under the most precarious conditions in relation to language, phrasing demands, appearing as a public or assembly, and being able to resist. For many, refusal is their only option, and the responsibility for creating alternatives (and reparations for past harms, see Palmer 2012) lies on the side of the community that harms them.

If humans change their attitude, and in the process of working towards more just multispecies communities, multispecies dialogues can include resistance or the kind of refusal that Honig describes. Sometimes dialogues between antagonistic communities are necessary to discuss how to avoid one another when sharing a habitat, or to draw other boundaries, but more intimate relations can also include periods of silence. 'No' is an answer that can end a conversation once and for all, or only temporarily. To become more attentive to the voices and forms of consent of others, we need to listen. As individuals, but, more importantly, also as societies.

Listening to others[2]

The knowledge structure that shapes political conversations and allows for certain voices to be heard while deeming others silent is not only characterized by a strong focus on human language, it also favors speech over silence, and speaking over listening (Bickford 1996). As we saw, understanding how to live better with nonhumans, and sometimes humans, does not simply mean we should invite others to be part of our conversations. Because the existing conversations are tailored to the capacities, interests and forms of expression of the class in charge, this could lead to exclusion on another level, and it would be patronizing too. Developing new conversations asks for reconsidering the framework of the dialogue itself, which begins with listening. Listening is not only important to hold meaningful conversations

2 Some of the following passages about listening appeared earlier in the article 'Deep listening and democracy' (2023).

with someone else: it can also create the conditions under which we can speak, through creating a common world (Bickford 1996) and opening up space for new, formerly unknown, voices to appear (Oliveros 1974). Because listening is other-regarding it matters not just in interpersonal relations, but can help to develop another political attitude too. This matters in times of ecological crises, but also with regard to the rise of the extreme right in many western countries (Latour 2018).

In political philosophy, including in theories of deliberation, listening is undertheorized (Morrell 2018). While it is described as necessary for successful deliberation, what listening entails and how we should listen is only explored by a few philosophers (Morrell 2018). In environmental deliberative philosophy, Andrew Dobson (2014) and John Dryzek (2000) emphasize the necessity of embodied listening to the more-than-human world in deliberating about it. Embodied listening matters because the natural world is not something outside of us, that we can understand solely with our minds, and write about from behind our computers. We are part of a larger living web, we think in part through our bodies, and we are always bodies in the world. To be able to understand and represent the world outside of us, we have to attend to it, and our body is part of this too. When we listen, we do not simply pick up information from the air with our ears: we are influenced by the air temperature, bacteria in our bodies, scents we might not register cognitively, emotions, and many other factors. Just as the whole body is part of speaking, the whole body is part of listening.

The most thorough analysis of the role of listening in deliberation is philosopher Susan Bickford's work on listening as a democratic practice (1996). Following Hannah Arendt, Bickford emphasizes the importance of plurality in political action, and of a common world in which political action can take place. All humans are different, and enter politics to defend their interests. For political action to take place, instead of just conflict, there needs to be a holding environment, a common world. According to Bickford, listening creates this space, in which people can show themselves to each other. This makes an agonistic form of collective judgement possible.

For Bickford, listening is both an embodied experience and a political stance. In describing that experience, she connects listening to attentiveness and to being still. In order to listen to someone, you have to take a step back. Bickford turns to Simone Weil, who wrote that 'all that is "I" must become passive in order to be attentive to the other; we have to become empty so that the other can enter' (1996, 144–5). But, according to Bickford, political listening also demands presence (1996, 144–5). To conceptualize this, she draws on the work of Merleau-Ponty. According to Merleau-Ponty, in political listening, the listener becomes the background against which the other can speak (1996, 146–147). The listener remains situated but directs their attention to the speaker, enabling them to step into the foreground. In a conversation, speaker and listener alternate roles.

This focus on the concrete other in the public arena is part of cultivating a universal political attitude. This creates the possibility of speaking with those who are different from us, and new forms of conversation. Bickford rightly draws attention to the importance of political listening in a world made up of many different perspectives. She also recognizes that listening is always situated and that power relations play a role in who is heard. However, her discussion of political listening is focused on those who are already seen as political agents. This is problematic, because it presupposes that the boundaries of the political community are fixed, while in fact they are always open to discussion and contestation. Moreover, beings who do not wish to be part of a human political community, such as some animal communities, also have political interests, sometimes interwoven with those of humans. Paying attention and listening politically to groups not recognized as citizens, or even as political agents, matters in order to discover what their interests are, how they relate to those of human political communities, and, in some cases, whether they should be recognized as part of the multispecies community. Political listening to these groups is important for reasons of justice, for epistemological reasons (we often do not know their position, which might affect ours) and because interests are entangled.

Transforming anthropocentric structures and foregrounding the perspectives of others raises many new questions regarding the borders of the political community, political speech and language, and multispecies justice. These cannot be answered by a philosopher outside of the public realm, but need to be answered following the process of listening itself, as some of the dialogues I discussed before show. It may be seem that listening to different more-than-human others is asking too much of us, given that humans already have considerable difficulty listening to other humans. But political listening is a skill that we need to cultivate towards all others. In the context of the current ecological crises, we need to learn to engage differently with earth others to even survive. In this process, Bickford's work offers a good starting point for listening to the voices of climate refugees, some children, and some nonhuman animals. But in other cases, her conception of listening and political communication is insufficient. We still know very little about many animal languages, for example, and animals speak not only through sounds, but also communicate through color, scent, and movement. Learning more about what they are saying and how they say it calls for new forms of listening. For certain human actors, too, such as those who do not speak or are very young, the conversations that Bickford describes are not possible or desirable. And there are always voices that are not yet heard, or not yet recognized as voices.

The concept of deep listening, developed by the American accordion player and composer Pauline Oliveros (1974), can help to think further about political listening to beings not yet recognized as political actors, whose language we do not understand – or about whom we do not even know whether they have a language – as well as to as-yet-unknown others. For Oliveros, deep listening is primarily a musical endeavor: it involves literally listening to music, sound rituals, improvisation, and environmental noises. Oliveros believes that we can learn to listen better and has developed compositions and exercises with that aim. The most well-known of these are her Sonic Meditations, a series of exercises or rituals that people can perform in groups. Sometimes these are very quiet, such as 'Native': 'Take a walk at night. Walk so silently that the bottoms of your feet become ears' (1974). In other cases, they are loud and require the use of loudspeakers, droning sounds, chanting, or yelling.

These sonic meditations can teach humans to listen better with their ears, but also to develop a different attitude to the world through listening, not dissimilar to what I called 'unselfing' at earlier points in this book, following Murdoch. In this sense, they are political meditations: Oliveros describes listening as a political practice and as a basis for forming judgements. Meditation is often associated with retreating inwards or stepping out of the world, but Oliveros shows that learning to listen better can help us to hear others better and perhaps even understand them. Recording technicians, composers, musicians or birdwatchers can hear more than most others, which shows us that we can all train to listen better, and become more present and attentive. But this literal meaning of listening can also be connected to its normative dimension. If I listen more carefully with my ears, I might learn to recognize the meaning of a bird's song in my garden. Through practising listening as a general attitude, I teach myself to become more open to the world of which I am part and what it has to tell me, and to give myself to that world.

Oliveros' listening offers more space for difference and the unknown than Bickford's. In deep listening, the listener becomes the background for an other who is not fixed and who may or may not appear, or may only appear after a long time. This is a different practice than Bickford's exchange of background and foreground by two beings who are similar to each other.

Many sonic meditations are intended to be practiced in groups, and aimed at collectively training ourselves in listening – this matters to cultivate new collectives and ways of performing community. Moreover, the world that is central to these meditations is not the human political world of Bickford or Arendt, but planet earth. Bodies, voices, musical instruments, footsteps, passing cars, birds, memories, forests, oceans, tape recordings, and other beings, events, and objects can be part of it.

Political listening in practice

Bickford's democratic listening and Oliveros' deep listening seem far removed from existing political discourse and the emphasis on certain forms of speech in contemporary politics. Yet within existing political practices and institutions, we can already find entry points for taking listening seriously. To investigate where and how we can start listening politically, we must first map out where political communication takes place and how different spheres of deliberation are connected. Political communication takes place in official political venues, such as in parliament or in regional or local councils; in counter-movements, such as demonstrations, social movements, citizen initiatives, and art; and in other spheres such as the media and universities. More-than-human political discussions are often found outside of spaces usually seen as political: in houses, nature reserves, cities, and parks. These various types of political communication have their own forms of political listening, so in order to develop these we need to be more specific than I can be here.

Generally, however, making more room for listening within existing political practices and institutions demands attention to time and space. Current political decision-making is strongly directed towards measurable outcomes. Meetings and assemblies are often clearly delineated in time. In parliament, for example, different points of view are presented, and then people vote. The listening here is often instrumental: the different parties involved are trying to convince others, and their standpoints are already decided upon. There is also little time for reflection on what is heard. Slowness is not a welcome guest in societies that are focused on efficiency. But a constant increase of pace comes at the expense of the strange, the new, and those who are othered and silenced. Holding several meetings would allow a clearer process to develop; the time in between creates a space in which listening can happen.

More time alone is not enough, however, for if participants keep repeating their own positions, this could result in less listening taking place. So, we also need to create new spaces in which those who make decisions listen to those affected by them. Listening sessions could take place in conference halls, hospitals, or in assisted living homes for young people; they could happen on farms, where both the farmers and the cows can be listened to regarding the transition to just food systems; or sessions could be held in nature reserves, whose inhabitants are negatively affected by pollution.[3] Listening well often (but not always) involves more than sitting quietly, and some activities, such as walking, lend themselves well to conversation and listening. Listening can also play a role in developing new collective political rituals.[4] Deep listening to the natural world can also be of importance in other forms of political decision-making, about nature reserves, seas, or other entities.[5]

On a small scale, there are initiatives to listen to others. Think for example of animal sanctuaries in which animal perspectives are taken seriously. Some of the proposals I discussed in relation to the North Sea in Chapter 5, such as the Sand Motor, also include listening. But introducing new formal spaces is necessary to open up the way for new forms of co-existing with others, as is learning the virtue of listening and silence as communities.

3 While political listening to other animals could take place on a farm or in a nature reserve, in these spaces there would still tend to be an unequal balance of power. Sue Donaldson (2020) draws attention to the importance of freedom of movement in improving multispecies relationships and communication. In order to find out what other animals want, humans need to literally give them more space, make the infrastructure safer, for example by banning cars, and create new public commons to which all kinds of animals have equal access and can encounter each other in freedom. In these commons, animals belonging to different species and communities could not only speak with each other in that space, but also about it. This asks us to be attentive to other forms of dialogue, which may involve interventions in the landscape – planting or not planting vegetation can structure conversations – learning about their languages, and using objects. For many companion animals, for example, food, toys, the sofa, and the car, already play an important role in communication with their humans.

4 Some indigenous communities, such as the Onondaga, begin their meetings by thanking the natural world that makes their existence possible. Kimmerer (2013) describes this as an expression of a world view in which humans are a part of a greater living whole, and as a way of recognising the agency of others in that web of being. Just like forms of greeting, which in politics can have the function of recognising one's interlocutor as a person, such rituals can make other forms of interaction with non-humans possible. Similar rituals could be developed in which listening to the nonhuman world has a central place. This could give humans new information about the larger world of which they are part and it would be a collective exercise in attention, a first step in learning an attitude in which humankind is not the master of others but part of a dynamic world with them.

5 When decisions are made about a natural entity or area, deep listening can be the starting point for deliberation. Human representatives could also play a part, perhaps because they possess specific knowledge about a particular situation or place, and listening can be formalized as part of the process of representation and learning about nature.

Political systems currently tend to magnify certain voices and silence others. Thinking about how those who express themselves differently can also have a voice in questions of common concern is of fundamental democratic importance. Simone Weil writes that for the public formation of opinion, freedom is not what matters most: there needs to be silence and attention in which the voices that are weak and frail can make themselves heard (Weil 1952). While more freedom matters greatly for developing new relations and conversations, especially with animals, Weil is right in stating it is not enough.

Learning to listen to others is a project that will never be finished, because there will always be new voices to listen too. More or better listening does not imply that everyone will want to speak, nor does it automatically lead to better collective judgments or understanding. There will always be beings who prefer to conceal themselves, some interests and ideas that cannot (yet) be expressed because of the form of language and dialogues, and no one knows where their words will end up in the future. In learning to listen, education can play a role, as can art and literature, but we also simply need to begin, in existing conversations and new ones. Like understanding, listening cannot be forced, but making space for it in political life and recognizing its importance can make new relationships possible. In the unknown there is always the possibility of something new, and better listening will make political discussions more difficult, beautiful, and rich.

Thank you for listening

The same is true for the conversations that I have with others and especially the ones I described in this book. They became not just more beautiful and difficult because I learned to listen better, but also because Olli, the mice, and the others, listened to me. Doris always listens to me, even when I do not notice it, and I do my best to listen to her as attentively as I can.

From the conversations that I described I not only learned about the individuals that I spoke with, but also about what it means to have a good dialogue with someone, and that the concept 'dialogue' can be extended to include new language-games. Further developing multispecies dialogues in and between communities asks for experiment, including creating spaces where we can have dialogues. Concepts like 'worldliness' and 'listening', 'language' and 'politics' can guide these conversations, as long as we critically review their history and present use. Other relevant concepts will likely come into view once we begin to act differently.

Viewing multispecies relations through the lens of dialogues shows that as humans, or philosophers, or animals, all of us are part of a bigger whole, in which nonhumans sometimes lead the way, like Olli and the mice did for me. Improving dialogues might sometimes be easy, when you meet someone who is willing to listen and shows good will, but it can also be hard work. Still, speaking with others and listening to them is an important part of understanding life and learning to live a good life, in a world we share with many others to whom we are connected, who are sometimes familiar and sometimes very unfamiliar.

So, now it is up to you, reader. Our dialogue has come to an end, but we might pick it up later, and perhaps the dialogues in this book will be part of new ones that you have with others, in the webs of speaking and listening that make up our shared world. I thank you for your time, because time is the most important gift we can give others, and look forward to your response.

Acknowledgements

In 2018 Bernice Bovenkerk invited me to become a postdoc in her project about animal agency in the Anthropocene. In a sense, this book is a response to the main research questions in that project: what is agency, and how can we think about agency and other concepts together with other animals, in a world so full of human domination. Bernice is also the first person who welcomed me into the world of animal philosophy when I was a student, and she read this manuscript very carefully, offering good and important suggestions. So, Bernice, I want to thank you three times, and I am grateful for your support throughout the years. My second thanks are also owed to someone who welcomed me into the world of the Dutch animal philosophy, and who read this manuscript with care and enthusiasm too: Clemens Driessen. Thank you, Clemens.

My thoughts in this book have also been shaped by ongoing dialogues with colleagues further away. First and foremost, by my ongoing conversation with Sue Donaldson. Thank you, Sue, for thinking together and for being my friend. Other colleagues that I have been fortunate to be in conversation with are Jessica Ullrich, Ralph Acampora, Elisa Aaltola, Will Kymlicka, Anat Pick, Angie Pepper, Dinesh Wadiwel, Danielle Celermajer, pattrice jones, Leonie Cornips, Josh Millburn, Alasdair Cochrane, Barbara Smuts, Steve Cooke, Emelia Quinn, Claire Jean Kim, Kari Driscoll, and the members of the Werkgroep Dierfilosofie, especially Brenda de Groot – thank you all. For academic and practical support, I want to thank Eloe Kingma, who always supports my animal (and other) projects at the University of Amsterdam, with care and a good sense of humor, and the other friendly people at ASCA, notably Esther Peeren. I also want to thank the NOW (Dutch Research Council) very much for funding my research in the past four years, and making the open access version of this book possible. And I want to thank the anonymous reviewers and journal editors who provided feedback on earlier versions of chapter 3, 8 and 9.

In addition to my living colleagues, I want to thank the dead philosophers with whose work I am in dialogue, especially Ludwig Wittgenstein, Jacques Derrida, Michel de Montaigne, Jean-Jacques Rousseau, Hannah Arendt and Zhuang Zi.

Thank you, Ilse, for being a reliable and communicative editor, for your faith in the project and your enthusiasm. Thanks to the other people at AUP too.

For ongoing conversations about other animals, plants and art, I want to thank my more-than-human human friends: thank you Gijsje, thank you Miek.

While the main insights in this book have been formed by the dialogues I had with all of my animal companions throughout the years – Putih, Pika, and Joy, but many others too – here I specifically want to thank the mice: Vachtje, Breedsnuit, Kleine muis, Mooitje, Grote muis, Flankie, Kraaloog, Witoog, Kleinoor, Bullie, Sneeuwtje, Stompie, Wolkje, Vlokkie, Madelief, Maanoor, Tweeteen, Mus, Spokie, Lieve, Roodmuis, Zeetje, Neushoorntje, Bram and Wezel. I also owe special thanks to Simba, Klontje and Kruidje, who now keep Doris and me company. And I want to thank the amphibians of Landsmeer, for teaching me to see the space we share differently and for being good neighbors.

I dedicate this book to Olli, who died when I was writing it, on April 7, 2023. He was my best friend for nearly ten years and I miss his jokes, gentleness and optimism, and that he always told me that things were alright. With him, they were.

I also dedicate this book to Doris, who watched over me when I was ill, kept me company when I could not bear the company of others, slept in the room while I was writing, walked with me, and loved me fiercely, which at times was the only thing I could hold on to.

References

Aaltola, Elisa. 2013. "Empathy, Intersubjectivity, and Animal Philosophy." *Environmental Philosophy* 10/2: 75–96.

Aaltola, Elisa. 2015. "Politico-Moral Apathy and Omnivore's Akrasia: Views from the Rationalist Tradition." *Politics and Animals* 1/1: 35–49.

Aaltola, Elisa. 2019. "Love and Animals: Simone Weil, Iris Murdoch and Attention as Love." In *The Routledge Handbook of Love in Philosophy*, edited by Adrienne M. Martin, 193–204. Routledge.

Acampora, Ralph. 2004. "Oikos and Domus: On Constructive Co-Habitation with Other Creatures." *Philosophy & Geography* 7(2): 219–35.

Acampora, Ralph. 2006. *Corporal Compassion: Animal Ethics and Philosophy of Body*. Pittsburgh: University of Pittsburgh Press.

Acampora, Ralph. 2021. "Zoögogy of the Oppressed." *Journal for Critical Animal Studies*, 1: 4–21.

Adamatzky, Andrew. 2022. "Language of Fungi Derived from their Electrical Spiking Activity." *Royal Society Open Science* 9/4, https://doi.org/10.1098/rsos.211926.

Adams, Carol. 2010. [1990] *The Sexual Politics Of Meat: A Feminist-Vegetarian Critical Theory*. New York: Continuum.

Adams, Carol and Josephine Donovan. 1995. *Animals and Women: Feminist Theoretical Explorations*. Durham: Duke University Press.

Adams, Carol J. and Josephine Donovan. 2007. *The Feminist Care Tradition in Animal Ethics: A Reader*. New York: Columbia University Press.

Adorno, Theodor W. (1984) "The Essay as Form." *New German Critique*, 32: 151–71.

Adorno, Theodor W. 1974. *Minima Moralia: Reflections from Damaged Life*. New York: New Left Books.

Adorno, Theodor W. 2000. *Problems of Moral Philosophy*. London: Polity Press.

Aho, Kevin. 2013. "Depression and Embodiment: Phenomenological Reflections on Motility, Affectivity, and Transcendence." *Medicine, Health Care and Philosophy* 16/4: 751–9.

Amann, Klaus. 1994. "Menschen, Mäuse und Fliegen. Eine Wissenssoziologische Analyse der Transformation von Organismen in Epistemische Objekte." *Zeitschrift für Soziologie* 23/1: 22–40, https://doi.org/10.1515/zfsoz-1994-0102.

Andrews, Kristin. 2020. "Naïve Normativity: The Social Foundation of Moral Cognition." *Journal of the American Philosophical Association* 6/1: 36–56.

Arendt, Hannah. 1998 [1958]. *The Human Condition*. Chicago: The University of Chicago Press.

Arendt, Hannah. 2021 [1977]. *Het leven van de geest*. Amsterdam: Ten Have.

Asenbaum, Hans. 2022a. "Doing Democratic Theory Democratically." *International Journal of Qualitative Methods*, 21: 1–12, https://doi.org/10.1177/16094069221105072.

Asenbaum, Hans. 2022b. "Democratic Theorizing." In *Research Methods in Deliberative Democracy*, edited by Selen A. Ercan, Hans Asenbaum, Nicole Curato, and Ricardo F. Mendonça, 66–79. Oxford: Oxford University Press.

Balcombe, Jonathan. 2016. *What a fish knows: The inner lives of our underwater cousins*. Scientific American/Farrar, Straus and Giroux.

Balcombe, Jonathan. 2022. "The Betrayed Fish: Reply to Oldfield." *Journal of Animal Ethics* 12/1: 59–62.

Bannon, Bryan E. 2009. "Animals, Language, and Life: Searching for Animal Attunement with Heidegger and Merleau-Ponty." *Environmental Philosophy* 6/1: 21–34.

Bekoff, Marc. 2002. *Minding Animals: Awareness, Emotions, and Heart*. Oxford: Oxford University Press.

Bekoff, Marc. 2007. *The Emotional Lives of Animals: A Leading Scientist Explores Animal Joy, Sorrow, and Empathy and Why They Matter.* Novato: New World Library.

Bekoff, Marc and Jessica Pierce. 2009. *Wild Justice: The Moral Lives of Animals.* Chicago: University of Chicago Press.

Bell Rizzolo, J. and G. Bradshaw. 2019. "Nonhuman Animal Nations: Transforming Conservation into Wildlife Self-Determination." *Society & Animals,* 29/4: 393–413.

Bennett, Jane. 2010. *Vibrant Matter: A Political Ecology of Things.* Durham: Duke University Press.

Bennett, Jane. 2020. *influx & efflux: writing up with Walt Whitman.* Durham: Duke University Press.

Bickford, Susan. 1996. *The Dissonance of Democracy: Listening, Conflict, and Citizenship.* Ithaca: Cornell University Press.

Biemann, U., and P. Tavares. 2014. *Forest Law/Selva Juridica: On the Cosmopolitics of Amazonia.* East Lansing: Eli and Edythe Broad Art Museum Michigan.

Blattner, Charlotte E., Kendra Coulter, and Will Kymlicka, eds. 2019. *Animal Labour: A New Frontier of Interspecies Justice?* Oxford: Oxford University Press.

Blattner, Charlotte E., Sue Donaldson, and Ryan Wilcox. 2020. "Animal Agency in Community: A Political Multispecies Ethnography of VINE Sanctuary." *Politics and Animals,* 6: 1–22.

Blissett, Sarah. 2021. "Algae Sympoiesis in Performance: Rendering-with Nonhuman Ecologies." *Performance Philosophy* 6/2: 117–36.

Botero, Maria. 2017. "Primates Are Touched by Your Concern: Touch, Emotion, and Social Cognition in Chimpanzees." In *The Routledge Handbook of Philosophy of Animal Minds,* edited by Andrews, Kristin, and Jacob Beck, 372–380. New York: Routledge.

Braithwaite, Victoria. 2010. *Do fish feel pain?.* Oxford: Oxford University Press.

Brown, Wendy. 2015. *Undoing the demos: Neoliberalism's stealth revolution.* Cambridge, MA: MIT Press.

Buchmann, Stephen L. 2023. *What a Bee Knows: Exploring the Thoughts, Memories, and Personalities of Bees.* Washington D.C.: Island Press.

Butler, Judith. 2002. *Gender trouble.* New York: Routledge.

Butler, Judith. 2015. *Notes toward a performative theory of assembly.* Harvard: Harvard University Press.

Butler, Judith. 2022. *The Force of Nonviolence: An Ethico-Political Bind.* London: Verso Books.

Calarco, Matthew. 2015. *Thinking Through Animals: Identity, Difference, Indistinction.* Stanford: Stanford University Press.

Calarco, Matthew. 2018. "The three ethologies." In *Exploring Animal Encounters,* edited by Ohrem, Dominik, and Matthew Calarco, 45–62. London: Palgrave Macmillan.

Camus, Albert. 2013 *The myth of Sisyphus.* London: Penguin UK.

Celermajer, Danielle, et al. 2020. "Justice through a Multispecies Lens." *Contemporary Political Theory,* 19: 475–512.

Celermajer, Danielle, et al. 2022. "Multispecies Justice: Theories, Challenges, and a Research Agenda for Environmental Politics." *Trajectories in Environmental Politics* 30/1–2: 116–37.

Cochrane, Alasdair. 2018. *Sentientist Politics: A Theory of Global Inter-Species Justice.* Oxford: Oxford University Press.

Cooke, Steve. 2021. "The Ethics of Touch and the Importance of Nonhuman Relationships in Animal Agriculture." *Journal of Agricultural and Environmental Ethics* 34/2: 1–20.

Cornips, Leonie. 2022. "The Animal Turn in Postcolonial (Socio) Linguistics: The Interspecies Greeting of the Dairy Cow." *Journal of Postcolonial Linguistics,* 6: 210–232.

Crary, Alice, and Lori Gruen. 2022. *Animal Crisis: A New Critical Theory.* Cambridge: Polity.

Cvetkovich, Ann. 2012. *Depression: A Public Feeling.* Durham: Duke University Press.

Dar, Anandini, and John Wall. (2011). "Children's political representation: The right to make a difference." *The International Journal of Children's Rights* 19(4): 595–612.

De Certeau, Michel. 2014. "Spatial practices." In *The People, Place, and Space Reader*, edited by Jen Jack Gieseking et al, 232–236. London: Routledge.
Derrida, Jacques. 2008. *The Animal That Therefore I Am*. New York: Fordham University Press.
Derrida, Jacques. 1978. "Cogito and the History of Madness." In *Writing and Difference*. Chicago: The University of Chicago Press.
Derrida, Jacques. 2005. *Sovereignties in Question: The Poetics of Paul Celan*. New York: Fordham University Press.
Descartes, René. 1985. [1637] "Letter to R. P. Vatier, 22 February 1638". In *The Philosophical Writings of Descartes, Vol. 3*. Cambridge. Cambridge University Press.
Descartes, René. 2013. *Meditations on First Philosophy*. Peterborough: Broadview Press.
Despret, Vinciane. 2016. *What Would Animals Say if We Asked the Right Questions?* Minneapolis: University of Minnesota Press.
De Ville, Jacques. 2010. "Madness and the Law: The Derrida/Foucault Debate Revisited." *Law and Critique*, 21: 17–37.
Diepenbeek, A. van, R. C.M. Creemers and J. J. C. W. van Delft. 2009. "Amfibieën en reptielen: Geschiedenis van de herpetologie in Nederland." *Natuur van Nederland* 9(1): 35–46.
Dobson, Andrew. 2014. *Listening for Democracy: Recognition, Representation, Reconciliation*. Oxford: Oxford University Press.
Dolensek, Nate, Daniel A. Gehrlach, Alexandra S. Klein, and Nadine Gogolla. 2020. "Facial Expressions of Emotion States and Their Neuronal Correlates in Mice." *Science* 368/ 6486: 89–94, https://doi.org/10.1126/science.aaz9468.
Donaldson, Sue. 2020. "Animal Agora: Animal Citizens and the Democratic Challenge." *Social Theory and Practice*, 46/4: 709–35, https://doi.org/10.5840/soctheorpract202061296.
Donaldson, Sue & Will Kymlicka. 2011. *Zoopolis: A political theory of animal rights*. Oxford: Oxford University Press.
Donaldson, Sue & Will Kymlicka. 2015. "Farmed Animal Sanctuaries: The Heart of the Movement?" *Politics and Animals* 1/1: 50–74.
Donaldson, Sue & Will Kymlicka. 2018. "Children and Animals." In *The Routledge Handbook of the Philosophy of Childhood and Children*, edited by Anca Gheaus, Gideon Calder, and Jurgen De Wispelaere, 282–93. Abingdon: Routledge.
Donovan, Josephine. 1990. "Animal Rights and Feminist Theory." *Signs: Journal of Women in Culture and Society*, 15/2: 350–75.
Donovan, Josephine. 2006. "Feminism and the Treatment of Animals: From Care to Dialogue." *Signs: Journal of Women in Culture and Society* 31/2: 305–29.
Driessen, Clemens. 2014. "Animal Deliberation." In *Animal Politics and Political Animals*, edited by David Schlosberg and Marcel Wissenburg, 90–104. London: Palgrave Macmillan.
Dryzek, John. 2000. *Deliberative Democracy and Beyond: Liberals, Critics, Contestations*. Oxford: Oxford University Press.
Dufourcq, Annabelle. 2014. "Is a World without Animals Possible?." *Environmental Philosophy* 11/1: 71–92.
Dufourcq, Annabelle. 2021. *The Imaginary of Animals*. New York: Routledge.
Duncan, Ian J. H. 2006. "The Changing Concept of Animal Sentience." *Applied Animal Behaviour Science* 100/1–2: 11–9.
Editorial Team. 2023. "Animals in Education: Ethical Perspectives." *On Education. Journal for Research and Debate* 6(16).
Emel, Jody, and Padini Nirmal. 2021. "A Feminist Research Agenda for Multispecies Justice." *A Research Agenda for Animal Geographies*, 23–37.
Escobar, Arturo. 2020. *Pluriversal Politics: The Real and the Possible*. Durham: Duke University Press.

Fisher, Mark. 2014. *Ghosts of my Life: Writings on Depression, Hauntology and Lost Futures.* London: John Hunt Publishing.

Foucault, Michel. 2010. [1975] *Discipline & Punish: The Birth of the Prison.* London: Penguin.

Foucault, Michel. 1998. *The History of Sexuality. Vol. 1: The Will to Knowledge.* London: Penguin.

Foucault, Michel. 2013. *History of Madness.* New York: Routledge.

Gaard, Greta. 2016. "Critical Ecofeminism: Interrogating 'Meat,''Species,'and 'Plant'." In *Meat Culture*, edited by Annie Potts, 264–87. Leiden: Brill.

George, Barbara, and Heather Manzo. 2022. "The Ohio River: Re-Imagining Water Risk through Embodied Deliberation." *Embodied Environmental Risk in Technical Communication.* Routledge, 99–118.

Geuss, Raymond. 2021. *A Philosopher Looks at Work.* Cambridge: Cambridge University Press.

Gibbens, Sarah. 2018. "These 'Sea Nomads' Are First Known Humans Genetically Adapted to Diving." National Geographic, https://www.nationalgeographic.com/science/article/bajau-sea-nomads-free-diving-spleen-science. Accessed October 1, 2024.

Gibson, Prudence, and Catriona Sandilands. 2021. "Introduction: Plant Performance." *Performance Philosophy* 6/2: 1–23.

Gillespie, Kathryn. 2018. *The Cow with Ear Tag #1389.* Chicago: University of Chicago Press.

Gordon, Deborah. 1992. "Wittgenstein and Ant-watching." *Biology and Philosophy* 7/1: 13–25.

Gruen, Lori. 2015. *Entangled Empathy: An Alternative Ethic for our Relationships with Animals.* New York: Lantern Books.

Habermas, Jurgen. "Three Normative Models of Democracy." *Constellations* 1/1: 1–10.

Haraway, Donna. 1991. *Simians, Cyborgs and Women: The Reinvention of Nature.* London: Routledge.

Haraway, Donna. 2003. *The companion species manifesto: Dogs, people, and significant otherness.* Cambridge: Prickly Paradigm Press.

Haraway, Donna. 2008. *When Species Meet.* Minneapolis: University of Minnesota Press.

Haraway, Donna. 2018. *Modest_Witness@Second_Millennium.FemaleMan©_Meets_OncoMouse™: Feminism and Technoscience.* London: Routledge.

Hall, Matthew. 2011. *Plants as Persons: A Philosophical Botany.* New York: Suny Press

Hasegawa, Tomohiro, and William B. Gudykunst. 1998. "Silence in Japan and the United States." *Journal of Cross-Cultural Psychology* 29/5: 668–84.

Haselmann, Lena, and Janke Klok. 2021. "Autumn Talk and Ocean Songs Dramatic Assemblage-Methods and Relevante of Perfomative Histography, Kultur, Kunst, Musik." *Musikgeschichte Auf Der Bühne. Performing Music History*, DOI:10.14361/9783839457467-021.

Hearne, Vicki. 2007. [1986] *Adam's Task: Calling Animals by Name.* New York: Skyhorse Publishing.

Heidegger, Martin. (1927). *Being and Time.* Malden: Blackwell Publishing.

Herbrechter, Stefan. 2023. "Unlearning to Be Human? The Pedagogical Implications of Twenty-first-century Post-anthropocentrism." *Posthumanism in Practice*: 212.

Herrmann, Kathrin, and Kimberley Jayne, eds. 2019. *Animal Experimentation: Working Towards a Paradigm Change.* Leiden: Brill.

Hobson, Kersty. 2013. "On the Making of the Environmental Citizen." *Environmental Politics* 22/1: 56–72.

Hoffmann, Roman, et al. 2022. "Climate Change Experiences Raise Environmental Concerns and Promote Green Voting." *Nature Climate Change* 12/2: 148–55.

Honig, Bonnie. 2017. *Public Things: Democracy in Disrepair.* New York: Fordham University Press.

Honig, Bonnie. 2021. *A Feminist Theory of Refusal.* Harvard: Harvard University Press.

Horta, Oscar. 2010. 'What is Speciesism?'. *Journal of Agricultural and Environmental Ethics*, 23/3: 243–66.

Howard, Len. 1952. *Birds as Individuals.* London: Collins Press.

Howard, Len. 1956. *Living with Birds*. London: Collins Press.
Hribal, Jason. 2010. *Fear of the Animal Planet: The Hidden History of Animal Resistance*. New York: AK Press.
Edmund Husserl. 1952. *Ideas Pertaining to a Pure Phenomenology and to a Phenomenological Philosophy, Second Book*. Dordrecht: Kluwer Academic Publishers.
Irigaray, Luce. 2011. "Perhaps Cultivating Touch Can Still Save Us." *SubStance* 40/3: 130–40.
Irvine, Leslie. 2001. "The Power of Play." *Anthrozoos: A Multidisciplinary Journal of The Interactions of People & Animals*, 14/3: 151–60.
Johnson, David. 2018. "Self in Nature, Nature in the Lifeworld." *Philosophy East and West* 68/4: 1134–54.
jones, pattrice. 2014. *The Oxen at the Intersection: A Collision*. New York: Lantern Books.
jones, pattrice. 2019. "Queer Eros in the Enchanted Forest: The Spirit of Stonewall as Sustainable Energy." *QED: A Journal in GLBTQ Worldmaking* 6/2: 76–82.
jones, pattrice. 2023. *Bird's-Eye Views: Queer Queries About Activism, Animals, and Identity*. Vermont: Vine Press.
Jones, Miriam. 2014. "Captivity in the Context of a Sanctuary for Formerly Farmed Animals." In *The Ethics of Captivity*, edited by Lori Gruen. Oxford: Oxford University Press, 90–101.
Kamphof, Ike. 2024. "Shapes of a Multispecies Lifeworld. Engaging with Amphibian Migrations." New Rituals for the Multispecies Community I Workshop, 16th May 2024.
Keulartz, Jozef, Henny Van Der Windt, and Jacques Swart. 2004. "Concepts of Nature As Communicative Devices: The Case of Dutch Nature Policy." *Environmental values* 13/1: 81–99.
Kim, Claire Jean. 2015. *Dangerous Crossings: Race, Species, and Nature in a Multicultural Age*. Cambridge: Cambridge University Press.
Kimmerer, Robin Wall. 2013. *Braiding Sweetgrass: Indigenous Wisdom, Scientific Knowledge and the Teachings of Plants*. New York: Milkweed Editions.
Ko, Aph and Syl Ko. 2017. *Aphro-Ism: Essays on Pop Culture, Feminism, and Black Veganism from Two Sisters*. New York: Lantern Books.
Kohn, Eduardo. 2013. *How Forests Think: Toward an Anthropology Beyond the Human*. Oakland: University of California Press.
Krause, Sharon R. 2019. *Freedom Beyond Sovereignty: Reconstructing Liberal Individualism*. Chicago: University of Chicago Press.
Kropotkin, Peter. 2021. *Mutual Aid: A Factor of Evolution*. Montreal: Black Rose Books Ltd.
Kusters, Wouter. 2020. *A Philosophy of Madness: The Experience of Psychotic Thinking*. Cambridge, MA: MIT Press.
Lamberti, P. 2023. *Befriending the Ears: The Transformative Power of Listening*. PhD Diss. University of Amsterdam.
Latour, Bruno. 2007. *Reassembling the Social: An Introduction To Actor-Network-Theory*. Oxford: Oxford University Pres.
Latour, Bruno. 2018. *Down to Earth: Politics in the New Climatic Regime*. London: John Wiley & Sons.
Lawrence, Anna M. 2022. "Listening to Plants: Conversations between Critical Plant Studies and Vegetal Geography." *Progress in Human Geography* 46/2: 629–51.
Lebra, Takie Sugiyama. 1987. "The Cultural Significance of Silence in Japanese Communication." *Multilingua*: 343–358.
Lemon, Alaina. 2015. "MetroDogs: The Heart in the Machine." *Journal of the Royal Anthropological Institute* 21/3: 660–79.
Li, Qing. 2019. *Into the Forest: How Trees Can Help You Find Health and Happiness*. London: Penguin UK.
Lundy, L. 2007. "The Lundy Model of Child Participation." *British Educational Research Journal*.

Mansbridge, Jane, James Bohman, Simone Chambers, Thomas Christiano, Archon Fung, John Parkinson, Dennis F. Thompson, and Mark E. Warren. 2012. "A Systemic Approach to Deliberative Democracy." In *Deliberative Systems: Deliberative Democracy at the Large Scale*, edited by John Parkinson and Jane Mansbridge, 1–26. Cambridge: Cambridge University Press.

Pye, Michael. 2014. *Aan de Rand van de Wereld. Hoe de Noordzee ons Vormde*. Amsterdam, De Bezige Bij.

McKinnon, Catriona. 2014. "Climate Change: Against Despair." *Ethics and the Environment* 19/1: 31–48.

Mancuso, Stefano. 2018. *The Revolutionary Genius of Plants: A New Understanding of Plant Intelligence and Behavior*. London: Simon and Schuster.

Marder, Michael. 2013. *Plant-thinking: A Philosophy of Vegetal Life*. New York: Columbia University Press.

McGuire, L., S. B. Palmer and N. S. Faber. 2022. "The Development of Speciesism: Age-Related Differences in the Moral View of Animals." *Social Psychological and Personality Science*: 1–10. https://doi.org/10.1177/19485506221086182.

McNiven, Ian. 2003. "Saltwater People: Spiritscapes, Maritime Rituals and the Archaeology of Australian Indigenous Seascapes." *World archaeology* 35/3: 329–349.

McNiven, Ian. 2016. "Sentient Sea: Seascapes as Spiritscapes." In *Handbook of Landscape Archaeology*, edited by David, Bruno, and Julian Thomas, 149–57. New York: Routledge.

Mediano, Pedro AM, Anthony Trewavas, and Paco Calvo. 2021. "Information and Integration in Plants: Towards a Quantitative Search for Plant Sentience." *Journal of Consciousness Studies* 28(1–2): 80–105.

Meijer, Eva. 2013. "Political Communication with Animals." *Humanimalia: A Journal of Human/Animal Interface Studies*, 5/1.

Meijer, Eva. 2014. "Stray philosophy: Human-dog observations on language, freedom and politics." *Journal for Critical Animal Studies* 12/4: 105–135.

Meijer, Eva. 2019. *When Animals Speak: Towards an Interspecies Democracy*. New York: New York University Press.

Meijer, Eva. 2020. "Stray Philosophy II: Dog/Human Reflections on Education, Boundaries, Care, and Forming Interspecies Communities." *Journal for Critical Animal Studies* 17/3: 3–30.

Meijer, Eva. 2021a. "Sanctuary Politics and the Borders of the Demos: A Comparison of Human and Nonhuman Animal Sanctuaries." *Krisis: Journal for contemporary philosophy* 41/2: 35–48, https://doi.org/10.21827/krisis.41.2.37174.

Meijer, Eva. 2021b. *The Limits of my Language. Meditations on Depression*. London: Pushkin Press.

Meijer, Eva. 2022a. "Learning Hope in the Anthropocene: The Party for the Animals and Hope as a Political Practice." *Animal Studies Journal* 11/1: 145–72.

Meijer, Eva. 2022b. "Learning to See Mice: (Stray Philosophy III)." *Humanimalia* 13/1: 203–51, https://doi.org/10.52537/humanimalia.11520.

Meijer, Eva. 2022c. *Misschien is een Ander Woord voor Hoop: Een Pleidooi voor Meerstemmigheid in het Politieke en Publieke Debat*. Amsterdam: De Geus.

Meijer, Eva. 2022d. *Verwar het niet met Afwezigheid: Over Politieke Stiltes*. Amsterdam: Uitgeverij Cossee.

Meijer, Eva. 2023a. "Deep Listening and Democracy: Political Listening to Fellow Citizens and Other Beings." *The Philosopher* 111/1: 53–60.

Meijer, Eva. 2023b. Developing Multispecies Education with Children and Animals. *on_education*: 16.

Meijer, Eva. 2024. "Tierkunst als Weltbildende Praxis." In *Nichtmenschliche Ästhetik: Kuratieren jenseits des Menschlichen*, edited by Jessica Ullrich, 39–50. Leiden: Springer.

Meijer, Eva. Forthcoming a. "The Meaning of Holes." In: *Animated Wor(l)ds*, edited by Elizabeth Tavella and Eva Spiegelhofer. Chicago: Chicago University Press.

Meijer, Eva. Forthcoming b. *Multispecies Assemblies*. Vermont: VINE Press.

Meijer, Eva, and Charlotte Blattner. 2021. "Animals and Climate Change." In *Justice and food security in a changing climate*, edited by Hanna Schübel and Ivo Wallimann-Helmer, 64–70. Wageningen: Wageningen Academics Publishers.

Meijer, Eva and Bernice Bovenkerk. 2021. "Taking Animal Perspectives into account in Animal Ethics." *Animals in Our Midst: The Challenges of Co-existing with Animals in the Anthropocene*, edited by Bernice Bovenkerk and Jozef Keulartz, 49–64. Leiden: Springer Science and Business Media Bv.

Melehy, Hassan. 2006. "Silencing the Animals: Montaigne, Descartes, and the Hyperbole of Reason." *symploke* 13/2: 263–82.

Merleau-Ponty, Maurice. 1962. *Phenomenology of Perception*. New York: Routledge.

Merleau-Ponty, Maurice. 1964. *Signs*. Evanston, IL: Northwestern University Press.

Merleau-Ponty, Maurice. 1968. *The Visible and the Invisible*. Evanston, IL: Northwestern University Press.

Merleau-Ponty, Maurice. 2003. *Nature: Course Notes from the Collège de France*. Evanston IL : Northwestern University Press.

Michelfelder, Diane. 2003. "Valuing Wildlife Populations in Urban Environments." *Journal of Social Philosophy* 34/1: 86.

Midgley, Mary. 1984. *Animals and Why they Matter*. Georgia: University of Georgia Press.

Midgley, Mary. 1985. "The Vulnerable World and Its Claims on Us." *The Georgia Review* 39/4: 739–754.

Monsó, Susana, and Birte Wrage. 2021. "Tactful Animals: How the Study of Touch can Inform the Animal Morality Debate." *Philosophical Psychology* 34/1: 1–27.

Monsó, S., Andrews, K., Doris, J. M., & Vargas, M. 2022. "Animal Moral Psychologies." In *The Oxford Handbook of Moral Psychology*, edited by Manuel Vargas and John Doris, 388–420. Oxford, Oxford University Press.

Montaigne, Michel de. 1958 [1595]. "Apology for Raymond Sebond." In *The Complete Works of Montaigne*. Stanford: Stanford University Press.

Montaigne, Michel. 2019. *The Essays*. London: Penguin UK.

Moore, Alison. 2019. "Using the Lundy Model of Child Participation with Adults." In *The Routledge International Handbook of Young Children's Rights*, edited by Jane Murray, Beth Blue Swadener, and Kylie Smith, 447–60. New York: Routledge.

Moore, Jason. 2016. *Anthropocene or Capitalocene?: Nature, History, and the Crisis of Capitalism*. Binghamton: Pm Press.

Moran S. 2022. "What Can Art Do for Ecological Thinking?" *The Ecological Citizen* 5/2: 103–8.

Morrell, Michael. 2018. "Listening and Deliberation." In *Oxford Handbook of Deliberative Democracy*, edited by André Bächtiger et al, 237–50. Oxford: Oxford University Press.

Morini, Marco. 2018. "Animals First!' The Rise of Animal Advocacy Parties in The EU: A New Party Family." *Contemporary Politics*, 24: 418–435.

Mortimer-Sandilands, Catriona, and Bruce Erickson (ed.). 2010. *Queer Ecologies: Sex, Nature, Politics, Desire*. Bloomington: Indiana University Press.

Murdoch, Iris. 1971. *The Sovereignty of Good*. London: Routledge.

Murdoch, Iris. 2003. *Metaphysics as a Guide to Morals*. London: Vintage Books.

Neimanis, Astrida. 2017. *Bodies of Water: Posthuman Feminist Phenomenology*. London: Bloomsbury Academic.

NOS. 2017. "Hondentekort in Nederland. " https://nos.nl/artikel/2166470-hondentekort-in-nederland-is-een-keurmerk-nodig. Accessed October 1, 2024.

NOS. 2021. "Meer Zwerfhonden uit het Buitenland en dus vaker Gedragsproblemen. " https://nos.nl/artikel/2394847-meer-zwerfhonden-uit-het-buitenland-en-dus-vaker-gedragsproblemen. Accessed October 1, 2024.
NIOZ. 2022. "Giving Land Back to the Sea, it Sounds Scary but Results in a Safer More Biodiverse Coast. " https://www.nioz.nl/en/news/giving-land-back-to-the-sea-it-sounds-scary-but-results-in-a-safer-more-biodiverse-coast. Accessed October 1, 2024.
NRC. 2022. "Watermanagers Roeren zich in Discussie. " https://www.nrc.nl/nieuws/2022/10/11/watermanagers-roeren-zich-in-discussie-we-moeten-ons-juist-niet-terugtrekken-voor-de-stijgende-zeespiegel-a4144835. Accessed October 1, 2024.
NVWA. 2020. *Zo doende 2019. Jaaroverzicht proefdieren en dierproeven.* Nederlandse Voedsel- en Warenautoriteit.
Oliver, Catherine. 2021. *Veganism, Archives, and Animals: Geographies of a Multispecies World.* Abingdon: Taylor & Francis.
Oliver, Kelly. 2009. *Animal Lessons: How They Teach Us To Be Human.* New York: Columbia University Press.
Oliveros, Pauline. 1974. *Sonic Meditations.* New York: Smith Publications.
Oswald, Freya, et al. 2020. "A Scoping Review Of The Relationship Between Running And Mental Health." *International Journal Of Environmental Research And Public Health* 17/21: 8059.
Pachirat, Timothy. 2011. *Every Twelve Seconds: Industrialized Slaughter and the Politics of Sight.* New Haven: Yale University Press.
Page, Joanna. 2020. "Planetary Art Beyond the Human: Rethinking Agency in the Anthropocene." *The Anthropocene Review* 7/3: 273–94.
Painter, Corinne. 2007. "Appropriating the Philosophies of Edmund Husserl and Edith Stein: Animal Psyche, Empathy, and Moral Subjectivity." In *Phenomenology and the Non-Human Animal: At the Limits of Experience,* edited by Painter, Corinne, and Christian Lotz, 97–115. Dordrecht: Springer Netherlands.
Palmer, Clare. 2012. "Can We – and Should We – Make Reparation to 'Nature'?" In *The Environment: Philosophy, Science, and Ethics,* edited by William P. Kabasenche, Michael O'Rourke, and Matthew H. Slater, 201–22. Cambridge: MIT Press.
Pearson, Chris. 2017. "History and Animal Agencies." In *The Oxford Handbook Of Animal Studies,* edited by Linda Kalof, 264–83. Oxford: Oxford University Press
Pederson, Helena. 2017. "Learning to Change the System from Within? Two Cases of Education for Animal Liberation." *Animal Liberation Currents.* http://www.animalliberationcurrents.com/learning-to-change-the-system/. Accessed October 1, 2024.
Peterson, Dale. 2012. *The Moral Lives Of Animals.* London: Bloomsbury Publishing.
Peterson, Debora, Natalia Hanazaki, and Paulo Cesar Simoes-Lopes. 2008. "Natural Resource Appropriation in Cooperative Artisanal Fishing Between Fishermen and Dolphins (Tursiops truncatus) in Laguna, Brazil." *Ocean & Coastal Management* 51/6: 469–75.
Pick, Anat. 2011. *Creaturely Poetics: Animality and Vulnerability in Literature and Film.* New York: Columbia University Press.
Pierce, Jessica. 2008. "Mice in the Sink: On the Expression of Empathy in Animals." *Environmental Philosophy* 5/1: 75–96.
Pleasants, Nigel. 2006. "Nonsense on Stilts? Wittgenstein, Ethics, and the Lives of Animals". *Inquiry, An Interdisciplinary Journal of Philosophy* 49/4: 314–36.
Plumwood, Val. 2002. "Prey to a Crocodile." *The Aisling Magazine.* http://www.aislingmagazine.com/aislingmagazine/articles/TAM30/ValPlumwood.html. Accessed October 1, 2024.
Plumwood, Val. 2001. "Nature as Agency and the Prospects for a Progressive Naturalism." *Capitalism Nature Socialism* 12/4: 3–32.

Plumwood, Val. 2014. "Nature in the active voice." In *The Handbook of Contemporary Animism*, edited by Graham Harvey, 441–53. New York: Routledge.
Probyn-Rapsey, Fiona. 2018. "Anthropocentrism." In *Critical Terms for Animal Studies*, edited by Lori Gruen, 47–63. Chicago: University of Chicago Press.
Robbins, William D., et al. 2006. "Ongoing Collapse of Coral-Reef Shark Populations." *Current Biology* 16/23: 2314–2319.
Rietveld, Erik. 2008. "Situated Normativity: The Normative Aspect of Embodied Cognition in Unreflective Action." *Mind* 117/468: 973–1001.
Rossello, Diego. 2022. "The Animal Condition in The Human Condition: Rethinking Arendt's Political Action Beyond the Human Species." *Contemporary Political Theory* 21/2: 219–39.
Sand Motor. https://dezandmotor.nl/en/. Accessed October 1, 2024.
Sandilands, Catriona. 2021. "Plant/s Matter." *Women's Studies* 50/8: 776–83.
Seeley, Tom. 2010. *Honeybee Democracy*. Princeton: Princeton University Press.
Segundo-Ortin, Miguel, and Paco Calvo. 2023. "Plant Sentience? Between Romanticism and Denial: Science." *Animal Sentience* 8/33: 1–33.
Simpson, Leanne Betasamosake. 2017. *As We Have Always Done: Indigenous Freedom Through Radical Resistance*. Minneapolis: University of Minnesota Press.
Simonsen, Kirsten. 2008. "Practice, Narrative and the Multicultural City: A Copenhagen Case." *European Urban and Regional Studies* 15/2: 145–158.
Smuts, Barbara. 2001. "Encounters with Animal Minds." *Journal of Consciousness Studies* 8/5–7: 293–309.
Soltis, Joseph, et al. 2014. "African Elephant Alarm Calls Distinguish between Threats from Humans and Bees." *PLoS One* 9.2: e89403.
Spivak, Gayatri Chakravorty. 1988. "Can the Subaltern Speak?" In *Marxism and the Interpretation of Culture*, edited by Cary Nelson and Lawrence Grossberg, 271–313. Champaign: University of Illinois Press.
Srinivasan, Krithika. 2022. "Re-animalising Wellbeing: Multispecies Justice after Development." *The Sociological Review* 70/2: 352–66.
Stickney, Jeff. 2020. "Philosophical Walks as Place-Based Environmental Education." *Journal of Philosophy of Education* 54/4: 1071–86.
TallBear, Kim. 2011. "Why Interspecies Thinking Needs Indigenous Standpoints." *Cultural Anthropology*. https://culanth.org/fieldsights/260-why-interspecies-thinking-needs-indigenous-standpoints. Accessed October 1, 2024.
Tanasescu, Mihnea. 2016. *Environment, Political Representation and the Challenge of Rights: Speaking for Nature*. Leiden: Springer.
Tanasescu, Mihnea. 2022. *Understanding The Rights of Nature: A Critical Introduction*. Bielefeld: transcript Verlag.
Tammi, T., R. Hohti, and P. Rautio. 2023. "From Child–Animal Relations to Multispecies Assemblages and Other-Than-Human Childhoods." *Barn – Forskning om Barn og Barndom i Norden*, 41: 2–3, https://doi.org/10.23865/barn.v41.5475.
Taylor, Affrica, and Veronica Pacini-Ketchabaw. 2018. *The Common Worlds of Children and Animals: Relational Ethics for Entangled Lives*. New York: Routledge.
Taylor, Sunaura. 2017. *Beasts of Burden: Animal and Disability Liberation*. New York: The New Press.
Tetsuro, Watsuji. 1961. *Climate and Culture: A Philosophical Study*. Westport: Greenwood Press.
Tiwari, Brijesh K., and Declan J. Troy. 2015. *Seaweed Sustainability–Food And Nonfood Applications*. Cambridge: Elsevier Academic Press.
Thomas, Evan. 2020. "Descartes on the Animal Within, and the Animals Without." *Canadian Journal of Philosophy* 50/8: 999–1014.

Toadvine, Ted. 2007. "'Strange Kinship': Merleau-Ponty on the Human–Animal Relation." In *Phenomenology of Life from the Animal Soul to the Human Mind: Book I In Search of Experience*, edited by Anna-Teresa Tymieniecka, 17–32. Leiden: Springer Netherlands.

Tschakert, Petra, et al. 2021. "Multispecies justice: Climate-just futures with, for and beyond humans." *Wiley Interdisciplinary Reviews: Climate Change* 12/2: e699.

Tsing, Anna Lowenhaupt. 2015. *The Mushroom at The End of the World: On the Possibility of Life in Capitalist Ruins*. Princeton: Princeton University Press.

Tsing, Anna Lowenhaupt. 2022. "The Sociality of Birds: Reflections on Ontological Edge Effects." In *Kin, Thinking with Deborah Bird Rose*, edited by Thom van Dooren and Matthew Chrulew, 15–32. Durham: Duke University Press.

Tuvel, Rebecca Dayna. 2014. *Epistemic Injustice Expanded: A Feminist, Animal Studies Approach*. PhD Diss., Vanderbilt University.

Ullrich, Jessica. 2019. "Animal Artistic Agency In Performative Interspecies Art In The Twenty-First Century." *Boletín de Arte,* 40: 69–83.

Ullrich, Jessica. 2022. "The Future of Art after the Animal Turn." *Animal Life and Human Culture Anthrozoology Studies,* 2022: 205–22.

Van Patter, Lauren E., and Charlotte Blattner. 2020. "Advancing Ethical Principles for Non-Invasive, Respectful Research with Nonhuman Animal Participants." *Society & Animals* 28.2: 171–190.

Vasseleu, Cathryn. 2022. *Textures of light: Vision and touch in Irigaray, Levinas and Merleau Ponty*. New York: Routledge.

Vasterling, Veronica. 2021. "Arendt's Post-dualist Approach to Nature: The Plurality of Animals." *HannahArendt.Net*. https://www.hannaharendt.net/index.php/han/article/view/461. Accessed October 1, 2024.

Venuta, Yamina. 2023. "Beyond the Mere Present: Husserl on the Temporality of Human and Animal Consciousness." *Continental Philosophy Review*: 1–17.

Vergani, Mario. 2021. "Husserl's Hesitant Attempts to Extend Personhood to Animals." *Husserl Studies* 37/1: 67–83.

Vincent, Sarah, Rebecca Ring, and Kristin Andrews. 2018. "Normative Practices of Other animals." In *Routledge Handbook on Moral Epistemology*, edited by Aaron Zimmerman, Karen Jones, and Mark Timmons, 57–83. New York: Routledge.

Vindas, Marco A., et al. 2016. "Brain Serotonergic Activation in Growth-Stunted Farmed Salmon: Adaption Versus Pathology." *Royal Society Open Science* 3/5: 160030.

Waal, Frans De. 2016. *Are We Smart Enough to Know How Smart Animals Are?*. New York: WW Norton & Company.

Wadiwel, Dinesh. 2015. *The War Against Animals*. Leiden: Brill Publishers.

Wadiwel, Dinesh. 2016. "Fish and Pain: The Politics of Doubt." *Animal Sentience: An Interdisciplinary Journal on Animal Feeling*, 1/3: 31.

Wadiwel, Dinesh. 2017. "Animal Friendship as a Way of Life: Sexuality, Petting and Interspecies Companionship." In *Foucault and Animals*, edited by Chrulew, Matthew, and Dinesh Joseph Wadiwel, 286–316. Leiden: Brill.

Wagenvoort, Hendruk. 1971. "Nehalennia and the Souls of the Dead." *Mnemosyne, Fourth Series,* 24/3: 273–92.

Wall, John. 2016. *Children's Rights: Today's Global Challenge*. Lanham: Rowman & Littlefield.

Weil, Simone. 2002. *Gravity and Grace*. New York: Routledge.

Weil, Simone. 1952. *The Need for Roots: Prelude to a Declaration of Duties Toward Mankind*. New York: Routledge Psychology Press.

Weisberg, Zipporah. 2009. "The Broken Promises of Monsters: Haraway, Animals and The Humanist Legacy." *Journal for Critical Animal Studies* 7/2: 22–62.

Weiss, Gail. 2006. "Can an Old Dog Learn New Tricks? Habitual Horizons in James, Bourdieu, and Merleau-Ponty." In *Intertwinings: Interdisciplinary Encounters with Merleau-Ponty*, edited by Gail Weiss, 223–40. New York: State University of New York Press.

Westling, Louise. 2013. *The Logos of The Living World: Merleau-Ponty, Animals, and Language*. New York: Fordham University Press.

Wittgenstein, Ludwig. 1958. *Philosophical Investigations*. London: Blackwell.

Wittgenstein, Ludwig. 1969. *Uber Gewißheit. On Certainty*. London: Blackwell.

Wittgenstein, Ludwig. 1978. *Lectures and Conversations on Aesthetics, Psychology and Religious Belief.* London: Blackwell.

Whiten, Andrew. 2021. "The burgeoning reach of animal culture." *Science* 372/6537, DOI: 10.1126/science.abe6514.

Wlodarczyk, Justyna. 2017. "Be More Dog: The Human–Canine Relationship in Contemporary Dog-Training Methodologies." *Performance Research* 22/2: 40–7.

Wohlleben, Peter. 2016. *The Hidden Life of Trees: What They Feel, How They Communicate Discoveries from a Secret World*. Vancouver: Greystone Books.

Wolch, Jennifer and Jody Emel. 1998. *Animal Geographies: Place, Politics, and Identity in the Nature-Culture Borderlands*. London: Verso Books.

Youatt, Rafi. 2017. "Anthropocentrism and the Politics of the Living." *Reflections on the Posthuman in International Relations*, https://www.e-ir.info/2017/10/18/anthropocentrism-and-the-politics-of-the-living/. Accessed October 1, 2024.

Young, Iris. 1990. *Justice and the Politics of Difference*. Princeton: Princeton University Press.

Young, Iris. 2000. *Inclusion and Democracy*. Oxford: Oxford University Press.

Young, Iris. 2005. *On Female Body Experience: "Throwing like a Girl" and Other Essays*. Oxford: Oxford University Press.

Zarobell, John. 2022. "Global Art Collectives and Exhibition Making." *Arts* 11/2, https://www.mdpi.com/2076-0752/11/2/38. Accessed October 1, 2024.

Zwamborn, Miek. 2020. *The Seaweed Collector's Handbook: From Purple Laver to Peacock's Tail*. London: Profile Books.

Zwamborn, Miek. 2022. "The Seawrack Gatherer." Dark Mountain Net. https://dark-mountain.net/the-seawrack-gatherer/. Accessed October 1, 2024.

Index

For Product Safety Concerns and Information please contact our EU representative GPSR@taylorandfrancis.com Taylor & Francis Verlag GmbH, Kaufingerstraße 24, 80331 München, Germany